Co-producing Knowledge for Sustainable Cities

At the current time, many issues and problems within sustainable urban development are managed within traditional disciplinary and organizational structures. However, problems such as climate change, resource constraints, poverty, and social tensions all exceed current compartmentalization of policy-making, administration, and knowledge production. This book provides a better understanding of how researchers and practitioners together can co-produce knowledge to better contribute to solving the complex challenges of reaching sustainable urban futures. It is written for academic and professional audiences working with urban planning and sustainable cities around the world.

Co-producing Knowledge for Sustainable Cities is presented, by way of introduction, as a non-linear, collaborative approach to knowledge production which combines interdisciplinary, transdisciplinary, and cross-sector approaches to societal problem solving. Examples are taken from five cities that are affiliated in different ways with Mistra Urban Futures, an international center for sustainable urban development. These cities include: Cape Town, Gothenburg, Greater Manchester, Kisumu, and Melbourne. Each city chapter discusses the drivers behind knowledge co-production and gives examples of activities and approaches that have been used to promote sustainable urban futures. Each chapter is written to promote mutual learning from the approaches that are already in use. Building upon these city cases, the conclusions outline topics for future research that can strengthen the promotion and implementation of the co-production of knowledge for urban sustainability across diverse urban contexts.

This book provides an overview of the diverse driving forces behind co-production, and their specific contexts and constraints in a variety of cosmopolitan cities. It outlines both the many added values of co-production, as well as the temporal, individual, and institutional problems within many co-production processes. The needs for normative reflection in processes that balance scientific excellence with the needs of societal change are exemplified through concrete examples. This book offers valuable lessons regarding the implications and potential impact that co-production processes can have for different user groups, such as planners, politicians, researchers, and NGOs in different urban development contexts.

Merritt Polk is an Associate Professor in Human Ecology, and current Head of Department at the School of Global Studies at the University of Gothenburg. Her main areas of research include differences in women's and men's travel patterns, gender mainstreaming in transport policy, the integration of sustainability in transport and urban planning, and transdisciplinary research methodology.

Routledge Research in Sustainable Urbanism

This series offers a forum for original and innovative research that engages with key debates and concepts in the field. Titles within the series range from empirical investigations to theoretical engagements, offering international perspectives and multidisciplinary dialogues across the social sciences.

Published

Co-producing Knowledge for Sustainable Cities
Joining forces for change
Edited by Merritt Polk

Forthcoming

Global Garbage: Urban Imaginaries of Waste
Excess, and abandonment
Edited by Christoph Lindner and Miriam Meissner

Co-producing Knowledge for Sustainable Cities

Joining forces for change

Edited by Merritt Polk

Routledge
Taylor & Francis Group
LONDON AND NEW YORK

First published 2015 by Routledge

2 Park Square, Milton Park, Abingdon, Oxfordshire OX14 4RN
52 Vanderbilt Avenue, New York, NY 10017

Routledge is an imprint of the Taylor & Francis Group, an informa business

First issued in paperback 2020

British Library Cataloguing in Publication Data
A catalogue record for this book is available from the British Library

Library of Congress Cataloging in Publication Data
Co-producing knowledge for sustainable cities : joining forces for change / edited by Merritt Polk.
pages cm. -- (Routledge research in sustainable urbanism)
Summary: "At the current time, many issues and problems within sustainable urban development are managed within traditional disciplinary and organizational structures. However, problems such as, climate change, resource constraints, poverty and social tensions all exceed current compartmentalization of policy-making, administration and knowledge production. This book provides a better understanding of how researchers and practitioners together can co-produce knowledge to better contribute to solving the complex challenges of reaching sustainable urban futures. It is written for academic and professional audiences working with urban planning and sustainable cities around the world. Co-producing Knowledge for Sustainable Cities is presented, by way of introduction, as a non-linear, collaborative approach to knowledge production which combines interdisciplinary, transdisciplinary and cross sector approaches to societal problem solving. Examples are taken from five cities including Cape Town, Gothenburg, Greater Manchester, Kisumu, and Melbourne. Each city chapter discusses the drivers behind knowledge co-production and gives examples of activities and approaches that have been used to promote sustainable urban futures. Each chapter is written to promote mutual learning from the approaches that are already in use. Building upon these city cases, the concluding chapter outlines topics for future research that can strengthen the promotion and implementation of the co-production of knowledge for urban sustainability across diverse urban contexts"-- Provided by publisher.
Includes bibliographical references and index.
1. Cities and towns--Study and teaching. 2. Sustainable urban development--Research. 3. Urban ecology (Sociology)--Research. 4. City planning. I. Polk, Merritt.
HT109.C67 2015
307.1'216--dc23
2014030167

ISBN: 978-1-138-81361-8 (hbk)
ISBN: 978-0-367-59979-9 (pbk)

Typeset in Times New Roman
by Taylor & Francis Books

Contents

List of illustrations

Figures

Tables

Boxes

List of contributors

Stig Egnell is a Swedish architect planner with long international experience. He shares his time as International Coordinator for Mistra Urban Futures and as Urban Adviser to Sida. He is owner of Stig Egnell Urban Management Consulting. His previous positions include Director for the City Planning Authority in Gothenburg and employment with VBB/SWECO, eventually as Head of the Department for Urban Planning and Design, member of VBB Board and Vice-president of SWECO. Stig has been resident in the Bahamas for UNDP and in Brazil as Manager for Sweconsult. Stig's educational background includes a Master of Architecture at Chalmers Technical University, Diploma from the Royal Academy of Fine Arts, School of Architecture and from the Nordic Institute for Studies in Urban and Regional Planning, Stockholm. His merits include prize-winning entries in several national and international planning competitions, the John Ericson medal from Chalmers University of Technology and the Aga Khan Award for Architecture in the Muslim World.

John Fien was Professor of Sustainability at RMIT University (2005–2013) where he was responsible for catalyzing cross-disciplinary sustainability research, and where he worked on projects concerned with youth leadership, sustainable consumption, environmental education, place-making, household environmentalism, organizational sustainability, and indigenous housing. He has been a member of the Australian National Commission for UNESCO and led research, policy development and capacity building projects for provincial and national agencies as well as for UNESCO, UNICEF, UNEP, WWF-International, the World Bank, and the OECD. John is currently Executive Director of the Swinburne Leadership Institute in Melbourne where his research is focused on leadership in relation to wicked problems and ways of enhancing leadership for the greater good, leadership development, and leadership in a greening economy.

Colin Fudge is President of RMIT Europe and Vice President of RMIT University. Professor Fudge has worked in universities and in government in the UK, Sweden, Australia, and in the European Commission. Previously he was Chair, European Union Expert Group on the Urban

Environment; Founding Director of the WHO Collaborative Research Centre on Healthy Cities and Urban Policy; Chair European Sustainable Cities and Towns Campaign; Visiting Professor, European University Institute, Italy, Tokyo University, Chalmers University and Royal Institute of Technology, Sweden; Member of the Commission for Architecture and the Built Environment (CABE) in England and Patron of the UK Urban Design Group. He has contributed through interdisciplinary and transdisciplinary research on public policy formulation and implementation; cities, sustainable development and adaptation to climate change; public health; demographic change and urban design. This has been recognized through his being awarded the Royal Professorship of Environmental Science by the Swedish Academy of Sciences and an Honorary Fellowship of the Royal Institute of British Architects.

Tomas Hellström is professor of Innovation, Entrepreneurship and Knowledge Creation at the School of Economics and Management, Lund University, Sweden. He has a Ph.D. in theory of science from Gothenburg University, and has previously worked at Chalmers University of Technology, Copenhagen Business School and University of Oslo. Tomas' main area of research is science policy, research evaluation and research management, but he has also published on methodology, innovation management, creativity and entrepreneurship. Previous work appears in journals such as *Research Policy*, *Human Relations*, *Science and Public Policy*, *Research Evaluation*, and *Geoforum*.

Jaan-Henrik Kain is an Associate Professor in Urban Transformation and Planning Processes at Chalmers Architecture, Chalmers University of Technology, Gothenburg, Sweden. Between 2010 and 2013, he was Research Director at Mistra Urban Futures. Jaan-Henrik's primary area of research is mechanisms that support or hinder sustainable urban development. The point of departure is to see cities/regions as constructed of interlinked urban structures (for example, transport networks, urban green, social space, housing) and where each such structure, in itself, is made up of various subsystems (such as institutions, physical structures, nature and users). As a researcher anchored in the fields of architecture and planning, he has been involved in numerous action-oriented research projects, involving facilitation of multi-stakeholder learning processes. Current research includes topics such as planning for urban ecosystem services, reviewing qualities, drivers and strategies for urban densification, in situ remediation of contaminated brown fields, and building democracy through waste management in development settings.

Mary Lawhon is a Lecturer in the Department of Geography, Geoinformatics and Meteorology at the University of Pretoria, focusing on urban political ecology in the African context. Her academic background combines natural and social sciences to provide an interdisciplinary understanding of the human-environment nexus. Specifically, she has used the cases of waste and alcohol regulation to examine broader questions of the politics of urban flows, informed largely by postcolonial theory and the move for theory from the global South. Mary holds a B.Sc. from the University of Kansas in environmental studies, a Master of Environment and Development from the University of KwaZulu-Natal, and a Ph.D. from Clark University. She recently completed a post-doctoral research fellowship at the University of Cape Town. She is a founding member of the Situated Urban Political Ecology collective.

Björn Malbert has a background of more than 15 years in urban planning practice. His work has been mainly directed at comprehensive and detailed physical planning practice and related research and development work, mainly for the Swedish Council for Building Research. As Professor in Design for Sustainable Urban Development at Chalmers Architecture, his research is directed at planning theory and methodology supported by knowledge development in close cooperation between research and practice. Björn's current research focus is on urban structures and questions related to everyday life in a sociotechnical perspective and following a systems analytic approach. His research concerns cities in industrialized as well as developing countries. A personal interest is directed at the roles and tools of professionals in participative planning processes for communication, mutual learning, decision-support and transdisciplinary knowledge production. Björn also co-coordinated the application writing process for the Mistra Urban Futures Center with Merritt Polk.

Tim May is Professor and Director of the Centre for Sustainable Urban and Regional Futures (SURF), University of Salford, Manchester, UK. Professor May's current work investigates the relationship between epistemic communities and communities of practice and explores values and learning in urban environments. Tim has authored, co-authored and edited 14 books (including new editions) that have been translated into 15 languages to date. He has recently signed contracts to produce two co-authored books with Beth Perry entitled *Reflexivity—A Guide* (Sage) and *Cities and Knowledge* (Routledge), as well as a third edition, with Zygmunt Bauman, of *Thinking Sociologically* (Wiley-Blackwell). Tim was the series editor of *Issues in Society*—an international book series with Open University Press/ McGraw-Hill—in which 17 books were published between 1998 and 2010. He has also written 80 journal articles and book chapters, edited 6 special editions of leading journals and authored over 100 research reports, government submissions, policy briefings and articles in different media.

Bernard Otieno Obera is an Urban Environmental Planning and Management practitioner. Ben has worked for several years for the Kisumu Municipal Council. As a Program Manager at Lake Victoria Region Local Authorities Cooperation (LVRLAC), Ben spearheaded the development of appropriate and competitive regional and country level programmes, establishing a knowledge management system for LVRLAC. As a Social Sector Specialist with the Millennium Cities Initiative, he was responsible for setting up the organization in Kenya, giving support to other cities, as well as developing and implementing projects towards the achievement of the Millennium Development Goals. During this period he helped reconvene the Kisumu Action Team post the 2007 election violence, which led to development of a reconstruction strategy, and formed the foundation for the Kisumu Urban Project supported by the French government. He has also been a researcher in the Kisumu Local Interaction Platform.

George Mark Onyango is an Associate Professor at the School of Planning and Architecture, Maseno University, Kenya. He has worked with various organizations across Africa in facilitating development interventions and building institutional capacities in various sectors. As a researcher and consultant, George has had practical experience as an expert with skills in both qualitative and quantitative approaches in planning, monitoring, and evaluation. As a professional, he has been actively engaged in building the knowledge and skills base of many development practitioners within and outside the university environment. His active engagement with the City of Kisumu was exemplified in participation in the Kisumu Action Team and later on the Kisumu Local Interaction Platform.

Zarina Patel holds a joint appointment with the African Centre for Cities and the Department of Environmental and Geographical Sciences at the University of Cape Town. Her research is focused on the political dimensions of policy and governance arrangements in pursuit of urban sustainable development. Uncovering of ways in which power is exercised through the use of specific and multiple knowledges and the range of value-based assumptions held by decision makers is a central strand of her research. A further strand traces the political geographies of environmental governance and the changing roles of the state and non-state actors in urban climate governance. Previous appointments include a senior lectureship in the School of Geography, Archaeology and Environmental Sciences at the University of Witwatersrand, and a British Academy Fellowship in the Geography Department at the University of Sheffield. She has sat on the boards of various environmental NGOs and statutory bodies in South Africa, and continues to serve on the editorial board of *Local Environment: International Journal of Justice and Sustainability*.

Beth Perry is Senior Research Fellow and Associate Director of the Centre for Sustainable Urban and Regional Futures, University of Salford Manchester, UK. Since 2011 she has been Director of the Greater Manchester Local Interaction Platform for Mistra Urban Futures. She is also co-investigator on a UK Arts and Humanities Research Council project on connecting communities in the creative urban economy. Beth's research focuses on the relationship between cities, governance and knowledge. She has edited four special editions of journals (*Regional Studies, Built Environment, Social Epistemology*, and *Sociology*) and authored 21 journal articles, 13 book chapters, and her work has been translated into French and Portuguese. She has collaborated with Tim May on *Social Research and Reflexivity* (2011) and the 4th edition of *Social Research: Issues, Methods and Process* (2011). Her most recent publications include a chapter titled 'Reflexivity and the Practice of Qualitative Research' (ed. Uwe Flick, *Sage Handbook of Qualitative Research*, 2013).

Merritt Polk is an Associate Professor in Human Ecology, and current Head of Department at the School of Global Studies at the University of Gothenburg, where she is a lecturer and researcher. She has worked actively with the establishment of Mistra Urban Futures, as a co-coordinator of the application writing process with Björn Malbert, and between 2010 and 2013, as Research Director. She has also been responsible for developing, testing and evaluating the transdisciplinary methodology at the Center since its inception. All of Merritt's research interests are practice based. She has worked closely with practitioners throughout her career in different topics within sustainable transportation and city planning. Her main areas of research include differences in women's and men's travel patterns, gender mainstreaming in transport policy, the integration of sustainability in transport and urban planning, and transdisciplinary research methodology.

Warren Smit has been a researcher at the African Centre for Cities at the University of Cape Town since 2008, and is the manager of the African Centre for Cities' CityLab program. He previously worked as a researcher on urban issues, from 1993 onwards, for the Urban Problems Research Unit at the University of Cape Town, the Built Environment Support Group at the University of Natal in Durban, and the Development Action Group in Cape Town. His research and policy work has involved extensive engagement with a wide range of government and civil society organizations and with academics from a diverse range of disciplines. His background is in urban planning, and his current main areas of research are urban governance, strategic urban planning, urban health and housing.

Acknowledgements

This book was made possible by the dedication and work of all of its co-authors. I would especially like to thank Tomas Hellström, Warren Smit, and Beth Perry for their supportive comments during the writing process, and Jaan-Henrik Kain for always stepping up to the plate whenever he was needed. The concluding chapter would not have been possible without the insightful input of the co-authors regarding their reflections on co-production from the vantage point of their respective cities. Thanks also to Simon Marvin and Tim May for their support in placing the book with an excellent publisher, and to Mistra Urban Futures who have funded the editing and also many of the contributions. Finally, I would like to thank all of the participants in the many different co-production processes and activities that not only make up the foundation of the work that is described in these pages, but that have inspired, challenged and cajoled us all into opening our minds and hearts to new ways of learning, communicating and doing research together.

List of abbreviations

ACC	African Centre for Cities
CAB	County Administration Board
CABE	Commission for Architecture and the Built Environment
CBK	Central Bank of Kenya
CBO	community-based organization
CBD	central business district
CDS	City Development Strategy
CSO	civil society organization
CWS	Cities without Slums
EPC	Environment Pedagogic Centre
ESRC	Economic and Social Research Council
GoK	Government of Kenya
GR	Gothenburg Region Association of Local Authorities
GMLIP	Greater Manchester Local Interaction Platform
ICRAF	International Centre for Research on Agroforestry
JKG	Jomo Kenyatta Grounds
JOOUST	Jaramogi Oginga Odinga University of Science and Technology
KANU	Kenya African National Union
KAT	Kisumu Action Team
KITREWA	Kisumu Town Residents Welfare Association
KLIP	Kisumu Local Interaction Platform
KTP	Knowledge Transfer Program
KUP	Kisumu Urban Project
LASDAP	Local Authorities Service Delivery Action Plan
LVBC	Lake Victoria Basin Commission
LVR	Lake Victoria Region
LVRLAC	Lake Victoria Region Local Authorities Cooperation
MCI	Millennium Cities Initiative
MCK	Municipal Council of Kisumu
MDG	Millennium Development Goals

MIIF	Manchester Innovation Investment Fund
MISTRA	The Swedish Foundation for Strategic Environmental Research
M: KC	Manchester: Knowledge Capital
MVP	Millennium Villages Project
NGO	Nongovernmental organizations
PPP	Public–Private Partnership
RDA	Regional Development Agency
RMIT	Royal Melbourne Institute of Technology
SDI	Slum Dwellers International
Sida	Swedish International Development Agency
SURF	Centre for Sustainable Urban and Regional Futures
UBC	Union of the Baltic Cities
UN	United Nations
UNDP	United Nations Development Program
VGR	Västra Götaland Region

1 Co-producing knowledge for sustainable urban futures

Merritt Polk and Jaan-Henrik Kain

Introduction

This book is about producing knowledge in urban areas. More specifically, it focuses on how decision-makers, civil servants, private actors and civil society come together with researchers to co-produce knowledge that can be used to contribute to creating more sustainable urban futures. This is by no means a new topic. What is new in the pages that follow is the breadth of examples of how different urban actors create knowledge together in cities around the world. Based upon first-hand experiences from researchers, planners, and officials working with urban development, we present concrete examples of projects, strategies, programs, and other activities that engage diverse voices and sources of expertise and experiences in Africa, Europe, and Southeast Asia. All of these examples show both the added values and critical issues that arise in knowledge production processes that attempt to bring together diverse actors, organizations, and spheres of activity to address current urban challenges.

From a historical perspective, serious environmental and social problems in cities are by no means a new phenomenon (Frank 2002; Hall 1988; Jacobs 1965). However, the scope and speed of change of current economic activities and resource consumption present hitherto un-faced challenges in understanding and dealing with the causal links that cut across continents, governing bodies, and value systems. Local governments are expected to provide services, promote economic and social development, and ensure safe and healthy environments within the constraints of short-term mandates, increasing social and environmental problems, and worsening financial conditions. International agendas connected to sustainable urban development similarly place formidable demands on cities not only to provide services locally, but also to promote participation in decision-making, environmental conservation, and social justice. While sustainable development and sustainability are well-established goals worldwide, different uses reflect a variety of underlying political worldviews and values, from market-driven mechanisms within neoliberal political agendas to ones that call for a total transformation of society based on social and environmental justice. This sets the stage for the discussion of sustainable cities that will be developed in this book.

What means do cities have at hand to meet such challenges and address such conflicts, both potential and actual? At the current time, cities are arenas for a variety of overlapping, mutually dependent, competing, and conflicting objectives. They are beset by essentially contested problems that are mired in uncertainty. They are the physical site for conflicts involving diverse value commitments from politics and bureaucracy, residents and business interests. While cities are managed within traditional organizational structures, many urban challenges such as climate change, resource constraints, poverty, poor health, and social tensions exceed the boundaries of the current compartmentalization of policy-making, planning, administration and academic knowledge production. Such challenges, by their nature, affect or engage a variety of stakeholders, decision-making levels, local–global contexts, disciplines, and sectors, as well as their respective social values, political ideologies, concrete urban conditions, and academic and bureaucratic structures. No single actor has the capacity or power to fully grasp or address this complexity. Nor is the knowledge needed to navigate the many challenges, uncertainties, and commitments readily available. It has to be created, or assembled, in dialogue between diverse users, political agendas, and interest groups. Creating spaces and processes that can support dialogue and collaboration, and complement existing governance and research capacity, is thus crucial to our ability to negotiate diverse value positions, capture relevant sources of knowledge and expertise, and better ground potential solutions in their specific contexts. Doing so is, however, far from simple.

This book highlights one approach to addressing the challenges facing all of us who are working with sustainable cities. It focuses on the need to include a wider range of actors in research processes in order to create novel forms of urban research that can better grasp the multidimensionality of the problems that are being faced. Sustainable development is a highly contested term. Consequently any discussion of sustainable cities must start from the perspective that sustainability means different things to different actor groups. It is therefore equally important for research on sustainable cities to be based on inclusive processes which can better capture the situated understandings of sustainability that exist in particular contexts by a variety of urban actors. We call this *knowledge co-production*. While co-production is used in many different areas and fields, here it is used to describe nonlinear, collaborative approaches to knowledge creation that draw upon interactive and participatory research approaches to societal problem solving. More specifically knowledge co-production refers to collaboratively based processes where different actors and interest groups come together with researchers to share and create knowledge that can be used to address the sustainability challenges being faced today, and increase the research capacity to contribute to societal problem-solving in the future. Knowledge co-production can thus be seen as a way to address the needs for democratic participation or more inclusive political processes called for within the sustainable development debate. For sustainable development to succeed, it needs to be grounded on social, political, and economic equity, but it needs greater equity in knowledge production processes as well. Thus the co-production approach developed here starts from the contested and inherently political nature of sustainable development, and shows how we can contribute to creating a more equitable political economy of knowledge production through different types of co-production processes.

Concrete examples of knowledge co-production are taken from Mistra Urban Futures, an international center for sustainable urban development based in Gothenburg, Sweden. The bulk of this book consists of five chapters written by practitioners and researchers from Gothenburg, Cape Town, Kisumu, Greater Manchester, and Melbourne. While the city of Melbourne is not officially affiliated with the center, as the other cities are, a researcher from Melbourne has been highly involved in the establishment of the center, and works with similar types of processes that add additional examples of relevant approaches to co-production. All of the city chapters describe and analyze different examples of co-production within their specific historical and social contexts. This includes researcher-practitioner-led projects in Gothenburg, a program for CityLabs in Cape Town, inclusive visioning processes in Kisumu, new relations between municipal governments and the university in Greater Manchester, and strategies for university–city engagement in Melbourne. As these examples show, at the center, knowledge co-production is viewed as essential for addressing complex challenges and problems in the urban environment. The work at the center thus addresses urban complexity through a focus on new modes of co-produced research that are used to promote mutual learning and action across diverse urban development contexts and conditions.

This introductory chapter lays the basis for why co-production is put forward as an important complement to traditional forms of knowledge production given current challenges and conditions of sustainable urban development. It starts by discussing the complexity of urban challenges and how current responses are most often compartmentalized and addressed within separate sectors, decision-making levels, and disciplines. We problematize how such traditional policy cultures and sector boundaries are closely linked to traditional models of knowledge production. We also discuss how these obstruct in-depth collaboration between policy-making, planning, and research, and limit genuine participation of societal interest groups from the private and civil sectors. All in all, this leads to a mismatch between critical urban challenges and the ability to develop research approaches that are tailored to address them. The penultimate section presents a more in-depth discussion of participatory approaches in planning and places co-production within the wider science–policy discussions. This includes the shift in planning from consensus and collaboration to more interactive research approaches. The chapter concludes with an outline of the remaining chapters in the book.

The challenges of sustainable cities

Promoting societal change in a sustainable direction is a topic that has been on policy-making, planning, and research agendas since the 1987 Brundtland commission report *Our Common Future* launched what became a global movement around sustainable development, and the work with Agenda 21 brought this global vision into local governance and research (United Nations 1992; WCED 1987). However, despite a close to thirty-year history, definitions and meanings connected to sustainability and sustainable development are still highly debated within policy, practice, and research. Different interpretations of sustainable development and their applications reflect a variety of underlying political and ideological worldviews and values (Kates *et al.* 2005; Owens and Cowell 2011; Robinson 2004). They show different degrees of connectivity between social activities, economic growth, and their resultant environmental impacts, and represent different beliefs about how such impacts are best addressed. Importantly, the importance of equity, especially the weight given to social justice issues at both global and local levels, varies greatly. To make things even more complicated, sustainable development is also interpreted and operationalized in conflicting and incompatible ways within the same local context by different political actors, business interests, and civil society organizations, as well as in different sectors and disciplines (Castro 2004; Robinson 2004).

Because of such differences in use, many have criticized the vagueness and ambiguity of sustainable development, arguing that the concept is purely rhetorical (Adams 2001; Castro 2004). Others describe this plurality of possible interpretations as *constructively ambiguous*, arguing that the "lack of definitional precision of the term sustainable development may represent an important political opportunity" (Robinson 2004: 374). In a similar vein, despite the many failures of sustainability and sustainable development to live up to their expected potential, their opportunistic use can also be seen to unveil the diverse value claims that exist in society, creating much needed arenas for political debate (Owens and Cowell 2011). Sustainable development can therefore be seen as not just a means for promoting a specific political agenda; it is equally a tool for exposing ideological beliefs, a space for conversation and collaboration, and a framework for creating incentives and participatory platforms for debate and action.

On top of such discursive challenges, applying the concept of sustainable development to cities also brings a number of other issues to the foreground. Reports on the millennium development goals show a number of improvements in urban areas around the world: fewer people living in extreme poverty, better access to clean water and sanitation, and a smaller proportion of slum dwellers around the world (United Nations 2012). However, important as they are, these gains mask deteriorating situations caused by the over-exploitation of habitats, global economic instability, climate change, and the failure of technocratic and market-based interventions to reduce emissions

(Griggs *et al.* 2013; IPCC 2014; Wikman and Rockström 2011). Social inequalities and misery continue to grow, while urban areas around the world are afflicted by severe environmental problems and resource degradation (Brugmann 2009; Davis 2006). Such trends show no sign of abatement. Despite the great amount of time and energy spent on promoting sustainability, for example in international and global goals and strategies, as well as sustainability projects and activities at grassroots and local levels, many would argue that we are no closer to reaching sustainability than we were twenty-five years ago (Griggs *et al.* 2013; UN-HABITAT 2010; UN 2013).

These problems are also likely to get much worse. The global population is expected to increase from today's almost 7 billion to 9.3 billion by 2050, with the greater part of this increase taking place in sub-Saharan Africa and parts of Asia (UN DESA 2010). As much as 70 percent of this global population is estimated to be urban. We are therefore experiencing one of the most significant changes in human history as global urbanization shifts from the North to the South (UN DESA 2009). There is a great risk of the current problems being further exacerbated in the future. In a speech at the EcoCities conference in 2011, Janice Perlman (2011), founder of the Mega-Cities project, maintained that "the entire growth of the world's population over the next thirty years is expected to be urban growth in the cities of the developing world, most of which will be concentrated in the informal sector in squatter settlements."

This prospect of massive urbanization processes largely materializing in informal settlements presents a challenge of historic proportions. Within this context, it is important to look at the interpretations of development that are being promoted under the umbrella of sustainable cities. Sustainable urban development is often equated with certain types of Western or Northern modernity, which is then applied to the rest of the world. However, questions such as "What kind of development?" and "Development for whom?" receive different responses from different geographic locations, civil sectors, development discourses, and ideological positions. Understanding the "urban now" thus involves conversations and interpretations spanning a diversity of local specificities, geographies, and histories (Robinson 2013). We therefore need to recast this traditional approach to modernity to "produce a conceptualization of modernity quite distinct from historical processes of modernization" (Robinson 2013: 660). This entails taking a critical look at the dominant development paradigms that promote Western versions of societal transformations within the debate on sustainable development (Castro 2004; Pieterse 2011; Robinson 2013). Consequently, knowledge about urban challenges and how to address them is best developed by engaging with the various practices, interactions, and rationalities of policy-makers, planners, developers, activists, and residents taking place in cities around the world (Watson 2003). Only then can such practice-based knowledge contribute to the development of more universally applicable theories on urban planning and practice (Parnell and Robinson 2012; Robinson 2013).

This is the foundation for the approach to knowledge co-production outlined in this book; namely, that the complexity of urban dynamics, rationalities, values, and power differentials cannot adequately be conceptualized or understood without a broad base of input. This in turn has repercussions for what sources and voices these interpretations come from and what legitimacy they hold. The way problems are framed determines their consequent analysis. The power to frame problems is thus also the power to determine what types and sources of knowledge are given legitimacy and relevance. Any discussion of sustainable cities must scrutinize the choices continually being made by politicians, municipal administrators, civil activists, the informal sector, business interests, and researchers. The phrase "knowledge is power" may be a cliché, but it does capture the crucial dynamics of urban relationships. Power produces different types of knowledge, and knowledge reinforces power; to put it simply, power and knowledge are entwined in the practice of knowledge production (Flyvbjerg 2001; Foucault 1980). The power to influence what is included in research questions, what methods are chosen and how knowledge is created is also the power of entitlement to partake in processes that have the potential to influence societal change.

Sector-based planning and decision-making

Another driving force behind the need for more inclusive forms of knowledge production are trends regarding the increasing fragmentation of cities into separate sectors, policy cultures, planning traditions, and decision-making levels. While comprehensive plans work across the city and sectors, day-to-day management, expertise, and decision-making end up increasingly based within distinct spatial and task-based units. For example, while traffic-flow and road-capacity calculations are done within transport authorities and administrations, residential and business development are the province of the city planning offices. This segmentation of thinking and organization becomes an additional barrier to sustainable urban development when the public sphere is divided into separate sectors, offices, commissions, councils, agencies, service providers, public works departments, and utility companies, and when service provision is increasingly being shifted into private and informal sectors. Dividing and distributing budgets according to mandate fulfillment and business opportunities leads to competition and territorialism, rather than to the cross-sector and cross-user integration needed to effectively move toward mutual long-term goals such as sustainable urban development.

At the same time as local governments are increasingly being compartmentalized into sectors and management units, globalization is resulting in an escalating connectivity where "everything is connected to everything else, and communication is anywhere, anytime" and "life becomes staggering in its possible complexity" (Batty 2012: 7). This pervasive connectivity generates unpredictable and rapid responses to any urban development initiative, where urban systems develop from the bottom up through messy, organic, discontinuous, and often chaotic processes: "Planned cities are always the exception rather than the rule and when directly planned, they only remain so for very short periods of time" (Batty 2009: 1045). This is nowhere more apparent than in the organic growth of informal settlements, where immediate survival needs, coupled with malfunctioning or corrupt governance, outweigh any attempts at planned growth.

Overall, the problems encountered when aiming to promote sustainable development in urban areas, tend to exceed current hierarchies and segmentations of policy-making, planning, administration, and knowledge production (Owens and Cowell 2011; Turok and Parnell 2009). The problems faced by cities are thus "radically cross-sector." Conditions in cities are also increasingly connected to environmental, social, and economic processes around the world, as climate change and the 2008 financial crises exemplify. Efforts to promote sustainable urban development must therefore manage a wide set of social, economic, and environmental global challenges and must do so within increasingly segmented and compartmentalized institutions and policy-making structures. However, urban planning and management often appear to be stuck with outdated, top-down intellectual traditions and praxis for framing problems and developing solutions, traditions that are no longer adequate for grasping or addressing complex problems and their discursive, spatial, and temporal complexity (Batty 2012). This is not surprising since the organizational structures from which we address current urban challenges have grown out of a complex web of historical and contextual factors. The legacy of colonialism, for example, includes city locations that were primarily designed to extract resources and service people on the other side of the world (Satterthwaite 2005).

From the linear to the interactive model of science

Different historical legacies also determine how scientific research has been organized within society. From a diverse history of science–policy interactions, today the most widespread relationship between science and the rest of society is *the linear model* (Godin 2006). In this model, knowledge is seen as moving from science to society. Science creates objective facts or truths that are used to support decision-making and action. Science claims to hold a certain authority or legitimacy regarding its ability to give a more accurate or objective picture of reality than non-scientific depictions. This scientific representation of reality is transferred to a policy context, where it is then translated as advice for specific policy decisions. In this way, science speaks "truth" to "power" (Jasanoff and Wynne 1998). As Pielke notes: "The linear model [also] suggests a role for scientists at the micro-level of particular policy contexts. It suggests that scientists simply need to get the science right in order to help improve decision-making in particular contexts" (2007: 86). While the linear model does present a coherent picture of the relationship between science and policy, it does so by ignoring some crucial characteristics of both science and policy-making. It also oversimplifies the interactions within and between science and policy.

To start with, there is no single or definitive "objective" or "scientific" view of reality; there are many. Science is a highly social activity. Individual scientists, with their diverse values and preferences, interact with disciplinary standards and methods in various scientific peer communities, creating a variety of approaches and interpretations even within the same discipline (Longino 1990). Furthermore, different disciplinary approaches and scientific interpretations, for example from the social and natural sciences, capture the diversity of the natural and social worlds through their specific disciplinary perspectives. Different disciplines describe and interpret different aspects of the world we live in, much like the blind men describe different parts of an elephant in the ancient Jain parable (Saxe 1872). Perhaps most importantly, scientists formulate their results primarily for their peer communities, not for decision-makers or the general public. There is therefore no obvious or unambiguous link between most scientific studies and their corresponding policy and decision-making contexts.

Despite such characteristics and issues, the linear model is still the most widespread the world over. It has a great deal of legitimacy among scientists and, perhaps more importantly, among non-scientists (Hulme 2009; Pielke 2007; Sarewitz 2004). One reason for its continued legitimacy is that it is highly applicable in situations where there is (apparent) agreement on the problem to be solved, and (presumed) certainty regarding the type of knowledge needed to solve it (Pielke 2007). However, as contestation and uncertainty increase, societal problems take on other, more complex characteristics (Funtowicz and Ravetz 1993). It is here that the linear model is unable to create certain and useful knowledge.

The best example of this dilemma is climate change. Here, debates from *both* science and policy stem from the linear model (Hulme 2009; Oreskes 2004; Sarewitz 2004). In the climate change controversy, debates rage over the interpretation, use, or nonuse of different scientific studies. Numerous accusations of bad science are made. Models and research are scrutinized and found wanting. Different scientific approaches are pitted against one another and vie for the premiere position of being the most applicable or true. The result is that both the quality and legitimacy of science are profoundly questioned in all of these exchanges.

As many researchers have pointed out, climate change is not only a scientific problem; it is also a political controversy (Hulme 2009; Oreskes 2004; Sarewitz 2004). More and better science is therefore not enough to resolve it. The climate change controversy reflects disagreements over political positions and interests regarding the environment, the use of oil, how to deal with uncertainty, and economic responsibility for future generations, to name but a few crucial topics (Hulme 2009; Oreskes 2004; Sarewitz 2004). It also reflects the diversity of legitimate methodological approaches from respective disciplines within the climate field (Bjurström and Polk 2011; Hulme 2009;

Oreskes 2004). While such ideological and cognitive disagreements may be irreconcilable, they do not indicate a lack of adequate science. Such cognitive disagreements mirror the plurality of viable scientific approaches, or blind men, who are engaged in describing the elephant of climate change. They also reflect a clear mismatch between the scope and complexity of socio-environmental problems, on the one hand, and the yearning for social or technical "quick fixes" on the other.

As this example shows, the linear model, in combination with disciplinary-based approaches to knowledge production, has serious weaknesses when it comes to addressing complex problems. New types of knowledge production, which are better able to negotiate contested issues and incorporate and process high degrees of uncertainty and complexity, are repeatedly called for. These calls have occurred most recently in the mid-1990s debates on the changing relationship between science and society, and have been referred to by a variety of terms, such as post-normal science, Mode 2, and transdisciplinary research (Funtowicz and Ravets 1993; Gibbons *et al.* 1994; Nowotny *et al.* 2001). Post-normal science (coined by Funtowicz and Ravets) and Mode 2 (coined by Gibbons *et al.*) refer to forms of knowledge production that exceed traditional, disciplinary-based scientific approaches and include, for example, multiple sites of knowledge production, higher degrees of contextualization, novel types of quality control and transdisciplinary approaches. *Transdisciplinary* is a concept that is often used in contrast to *multidisciplinary* (where disciplines work side by side on a common theme) and *interdisciplinary* (where a certain degree of integration occurs between disciplines in their problem formulations, and use of methods and/or theories) (Frodeman 2010; Klein 1990). Transdisciplinary is used to refer to both the transcending of disciplines (a higher degree of interdisciplinary integration), and to the transcending of the boundaries of the university itself (Frodeman 2010; Klein 1990). In this book transdisciplinary will be used to describe knowledge production that not only integrates knowledge from different disciplines (interdisciplinary), but also includes values, knowledge, know-how, and expertise from non-scientific sources, the second use above (Klein 2010). The approach to knowledge co-production taken in this book can therefore be seen as a type of transdisciplinary research.

With these debates as a backdrop, we can currently see two major and somewhat contradictory, parallel trends at play within university organizations and institutions. In the first trend, the division of labor into traditional subject categories and disciplinary approaches still typifies the intellectual and organizational structures of many knowledge-producing institutions (Frodeman 2010; Klein 1990). This shows the extent to which universities are still predominantly modeled and structured along disciplinary lines, despite the increasing number of calls for interdisciplinary and transdisciplinary research (Gibbons *et al.* 1994; Klein 1990; Weingart 2010). The entire system of incentives, promotions, teaching, and funding within universities is predominantly based upon disciplinary peer review, and in this milieu interdisciplinary and transdisciplinary approaches struggle to establish alternative standards of quality control and legitimacy (Holbrook 2010; Huutoniemi 2010; Russell *et al.* 2008).

The second trend is manifested in the explosion of university organizations and research approaches that are working actively with different aspects of sustainability and sustainable development and are attempting to bridge the science–policy gap through more collaborative modes of knowledge production (Gibbons *et al.* 1994; König 2013; Nowotny *et al.* 2001; Trencher *et al.* 2013). This trend has grown out of the coevolution of science and society and the increasing pressures on science to contribute to societal development, and especially to economic growth (Gibbons *et al.* 1994; König 2013; Nowotny *et al.* 2001; Trencher *et al.* 2013). Here, in addition to the first mission of universities (teaching) and the second mission (research), universities are engaging more and more in what since the late 1980s has been referred to as a *third* mission, which includes different types of societal engagement (Boyer 1996; Venditti *et al.* 2013). However, while purporting to support new modes of knowledge production, recent studies show that many examples of the third mission are still based on a linear model of science and innovation (Russell *et al.* 2008; Venditti *et al.* 2013). Thus, despite calls for new relationships between science and society and claims that a plurality of voices is needed, new scientific roles continue to develop within traditional approaches to science–policy relations (Russell *et al.* 2008; Turnhout *et al.* 2013; Venditti *et al.* 2013).

As these examples show, the relationship between universities and society has a complicated and heterogeneous history, one that is continually evolving in response to societal and economic incentives and to pressures from within and outside of university cultures (Nowotny *et al.* 2001). Current knowledge-producing organizations, such as universities and consulting firms, are very successful at producing high-quality knowledge that focuses on specific sectors and disciplinary areas. While this type of knowledge production leads to excellent results with certain types of problems, the cross-sector, multilevel, power-differentiated complexities currently facing urban practitioners are outside of its grasp. Consequently, while universities could be central actors in such knowledge production they often fail to serve that function. Neither universities nor research funding agencies are designed to support boundary-crossing modes of knowledge production that address and negotiate conflicting social interests and can respond to problems that transgress both disciplinary and organizational boundaries. Since the production of such socially robust knowledge and learning demands active engagement from both science and social actors, there are currently few clear sites for their methodological and theoretical development within traditional university settings (König 2013).

From consensus and collaboration to knowledge co-production

The focus on knowledge co-production taken in this book attempts to address some of the limitations outlined above regarding the multiple framings given to sustainability, the segmentation of planning and practice, the complexity of current urban challenges and the dominance of the linear model of science–policy relations. The need for different types of knowledge in planning has a long history of debate. During the 1950s and 1960s, for example, urban planners and policy-makers typically assumed that the complex behavior of large cities could be predicted by experts and then properly planned and managed through so-called rational or synoptic top-down policy and action (Hudson 1979; Jackson 1991). The "rationale for planning within modernism is that knowledge can be harnessed through planning to achieve positive change. The notion of progress is inherent to modernism so that as knowledge accretes over time, societal improvement follows from the use of more and better knowledge through planning" (Rydin 2007: 53). However, planning based on a hierarchical approach was not successful in responding to major challenges, such as housing and social inequalities (Rosenhead 1989). In the mid 1960s, for example, Jane Jacobs (1965) described how the rational planning models of the time failed to achieve livable cities in the US.

There are numerous examples where both planning studies and science in general have shifted away from seeing knowledge as an "objective truth" that can be created within university settings or by experts of different kinds and then applied in practice, toward understanding knowledge as a more dynamic and multifaceted concept embedded in different social processes (Nowotny *et al.* 2001; Rydin 2007; Watson 2003). In planning, this shift developed into what is referred to as the "communicative turn" which emphasizes dialogue and consensus-building as the foundation for attaining positive urban change (Healey 1992). However, communicative planning was increasingly criticized for being naive because existing institutional arrangements and prevailing power relationships clearly obstructed the ideal of "undistorted communication" (Huxley 2000; Lauria 2000).

To address such power asymmetries and institutional inadequacies, planning theorists have continued to examine communicative planning from different perspectives. John Forester (1989), for example, acknowledges that consensus is not always possible due to persistent asymmetrical power relations in planning situations. Judith Innes (1996) argues for mediated processes of group deliberation and negotiation for consensus-building to bring all types of interests into the conversation. Patsy Healey (1997) emphasizes the need for institutional structures to support and secure undistorted communication among diverse stakeholders. Sandercock and Forsyth (1992) prioritize emancipation and empowerment rather than consensus to accommodate diversity and to challenge inequalities.

While all of these approaches to communicative planning bring up crucial aspects of participation in planning, a majority (if not all) are based on the planning contexts and urban development issues of the liberal democracies and more or less mature economies of the global North (Watson 2014). Following the myth of "modernization," it has been assumed that a "transnational flow of planning ideas and practices" toward the South supports benevolent urban development paths (Healey 2012: 188). However, in reality, each planning context is permeated by local specificity, contingencies, and complexity (Robinson 2011, 2013). The diversity and local specificity of knowledge means that the role of the planner is challenged as urban knowledge becomes more distributed and collaboratively produced. In consequence, recent empirical and theoretical studies from the global South raise wider sets of concerns including a continuing critique of mainstream development, and lack of empirical and theoretical cases and insight from non-Western origins (Watson 2014). To understand the "urban now" (Robinson 2013), planners need to listen to unheard voices and manage multiple knowledges through deliberative and collaborative approaches, regardless of whether these knowledges are lay, local, experiential, intuitive, or embodied (Rydin 2007: 55). In doing so it is critical to distinguish knowledge claims from other (and potentially equally important) claims rooted in ethics, pragmatism, aesthetics, rationalities, or power (Rydin 2007; Watson 2003). This means that planners first need to recognize and also listen to the less powerful stakeholders—*opening up* the knowledge production process—and then be capable of testing the relevance of knowledge claims—*closing down* the same process (Rydin 2007).

Such a revamping of planning has many important implications. However, in this book we focus on just one of them, namely different approaches to joint knowledge production. The term *co-production* is used to capture a diversity of collaborative processes where different urban actors share knowledge, experiences, and expertise to create solutions for societal problems. Co-production is a concept that has been used to describe a variety of interactions between diverse social groups. It occurs in a number of subjects including urban studies, natural resource management, development studies, sustainability science, and transdisciplinary research (Albrechts 2012; Mitlin 2008; Pohl *et al.* 2010; Robinson and Tansey 2006; Watson 2014). Though they comprise a variety of different approaches, what all of these uses have in common is a concern with solving complex societal problems by integrating experiences and knowledge from diverse groups in society.

In a recent literature review and discussion article, Vanessa Watson (2014) examines the co-production of collective goods and services by different actor groups in urban settings. She distinguishes between two main categories: state-initiated and social-movement-initiated co-production. The distinguishing characteristic of *state-initiated* co-production according to Watson is that the state instigates and retains power over the co-production process (Watson 2014: 65). A majority of the examples she provides include sharing responsibility with nongovernmental actors for cost-effective service provision, for example in crime reduction, sanitation systems, and education (Ostrom 1996; Watson 2014). The second category, *social-movement-initiated* co-production, shifts the balance of power into the hands of civil society and the informal sector. This is exemplified through the work done by Shack/Slum Dwellers International (SDI) and includes strategies that aim at empowering citizens and adapting governance to fit their needs, often through the work of nongovernmental organizations (NGOs) (Mitlin 2008; Watson 2014). Examples of social-movement-initiated co-production include self-enumeration and mapping, joint planning and upgrading of neighborhoods, and innovative trans-local learning and knowledge exchanges (Watson 2014).

It will be argued here that the new modes of bridging the science–policy gap discussed above provide two additional categories of co-production that can be added to Watson's typology (see Figure 1.1). First, the increasing direct engagement of universities in improving the living conditions and livability of our cities positions them as a particular case of state-initiated co-production of urban knowledge. This will be referred to here as *researcher-initiated co-production*. The defining feature of researcher-initiated co-production is its origin in research and educational institutions and organizations. Here the focus is more specifically on combining scientific knowledge production with other forms and sources of knowledge and expertise to produce different types of knowledge products that can feed back into both practice-based and scientific knowledge production. This category includes a variety of methods, approaches, and degrees of knowledge exchange and integration between the spheres of science and practice. Some examples are: interactive social science, issue-driven interdisciplinarity, sustainability science, and transdisciplinary research (Caswell and Shove 2000; Robinson 2008; Robinson and Tansey 2006; Talwar *et al.* 2011; Wiek *et al.* 2012; Wiek *et al.* 2014; Zierhofer and Burger 2001).

Second, as research policies are increasingly oriented toward the third mission of universities and the integration of research and innovation to promote economic development, various business sectors continue to play a decisive

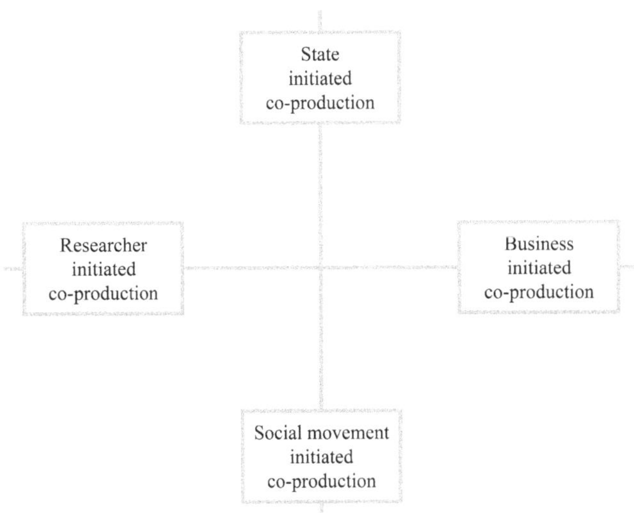

Figure 1.1 Different categories of co-production
Source: Developed from Watson 2014.

role in urban development and an increasing role in sustainable urban development. One example is the triple helix model, which is used to describe and analyze the role of universities in promoting social and economic development (Etzkowitz and Leydesdorff 2000; Hessels and van Lente 2008). This model illustrates specific examples of university–industry–government relations in knowledge production (Leydesdorff and Meyer 2006) and how university research becomes linked to corporate funding and commercialization (Russell *et al.* 2008). While triple helix approaches are initiated from both state and research organizations, business interests are more and more taking the lead in initiating different types of research as well. A final category of co-production can therefore be labeled *business-initiated co-production*. Here the goals of knowledge production focus on innovations and research that can be commercialized and translated into different types of intellectual property rights and marketable products.

While we distinguish between different initiators of co-production, such categorizations do not capture the full complexity of the relationships and processes that exist in society. Often, one or more types of organizations and institutions initiate co-production processes together, or there are varying degrees of participation of different actors in different phases. A final distinction can therefore be made regarding the outputs and aims of co-production processes that will lend more clarity to the differences between them. To a great extent, the origins determine who has control of the process, as noted by Watson above. This includes the specific goals of the co-production processes, for example, whether it is focused on the delivery of goods and services, producing knowledge products (scientific publications, policy input, reports) and/or creating marketable products, or a combination of all of these. As Watson concludes in her article, and as is also discussed in Mitlin, while the goal of state and social-movement-initiated co-production is collective goods and services, they argue that such processes importantly also result in increased emancipation, capacity-building, and political consolidation (Mitlin 2008; Watson 2014). In researcher-initiated co-production, the outputs of co-production have a dual focus as well, namely the production of scientific knowledge and input that can feed into both policy-making and societal change processes, as well as into scientific knowledge production. While co-production initiated by state, social movement, and business actors can also include combining scientific knowledge production with other forms and sources of knowledge and expertise, they do not usually produce scientific knowledge as an output. A distinguishing characteristic of researcher-initiated co-production is therefore the dual goals of producing both scientific knowledge and usable outputs for societal change. In many examples of researcher-initiated co-production, researchers maintain control of the knowledge production processes. However, there are a growing number of more symmetrical and jointly owned and led processes that attempt to share responsibility for knowledge production and to create knowledge that is both socially relevant and scientifically reliable (Hirsch Hadorn *et al.* 2008; Lang *et al.* 2012; Wiek *et al.* 2014).

All of the different approaches to co-production outlined above have a great deal of merit and are deserving of further attention. However, given the issues outlined in this chapter regarding the challenges facing urban practitioners and the prevalence of the linear model of science–policy relations, the overall focus of this book is on how researcher-initiated co-production can contribute to sustainable cities. This includes a number of different interdisciplinary and transdisciplinary approaches from urban studies, geography, architecture, human ecology, political science, and sociology. As will be clear in the following chapters, state, social movements, and business organizations have played important roles in many of the examples that are given, especially in the early stages. As we will also show, openness to novel ways of working together develops historically over time, resulting in different degrees of interaction between a variety of societal actors. However, what a majority of the examples in the following chapters share is an interest in combining scientific perspectives with other types of knowledge and expertise, and reintegrating the results back into both scientific and practice-based contexts.

Outline of the book: examples of knowledge co-production in diverse urban contexts

The remainder of this book centers on in-depth descriptions of co-production from a number of urban development contexts. These include Gothenburg, Cape Town, Kisumu, Manchester, and Melbourne, as well as a number of examples from cities in Southeast Asia. Each city case is situated in its historical context including highlights of the particular social, environmental, political, and economic trends that create and influence current challenges. Each chapter describes a number of concrete examples of knowledge co-production, and concludes with a summary of the experiences of co-production from their specific locations.

While the focus is on examples of co-production initiated by research organizations, we also give examples initiated by the state, social movements, and businesses, as well as different combinations of these categories. Though all approaches share an overall common goal of promoting sustainability, the various trends in the cities have grown from different constellations of ideas and problems, with distinct local challenges. The different governmental and research institutions that are represented here have developed in response to a specific set of barriers and challenges in the cities themselves as well as in accordance with particular planning and disciplinary/interdisciplinary approaches. Each city case is therefore viewed through a particular lens composed of practice and research approaches.

In Chapter 2, Björn Malbert, Merritt Polk, and Stig Egnell outline the history of the motivations and drivers behind the needs for collaboration between public and academic organizations in Gothenburg, Sweden. Starting with a brief history of recent urban development in the city, from the shipyard crash in the 1970s, to the new city-branding strategies of the 1990s, this chapter describes some of the different attempts to link urban development more tightly with knowledge production through collaboration between public, private, and university actors. It describes the conditions and constraints of recent co-production processes by examining a number of strategies and activities that have been carried out around key local and regional urban challenges, including regional networks for sustained growth, cross-sector programs for public transport, activities for social sustainability, policy-based dialogue processes, and researcher-initiated co-production. With its descriptions of the history and structure of Mistra Urban Futures, this chapter sets the stage for the rest of the book.

In Chapter 3, Warren Smit, Mary Lawhon, and Zarina Patel reflect upon the experience of co-producing knowledge at the African Center for Cities (ACC) in Cape Town, South Africa. The ACC emerged out of a long history of practical collaboration in South Africa between government, civil society, and academics, and now exists as an institution that seeks to build knowledge in conjunction with practitioners. The chapter discusses the CityLab program, which consists of a number of interdisciplinary platforms dedicated to thematic research areas based in and focusing on Cape Town. It also presents the more recent Knowledge Transfer Program, a collaboration between the ACC and the City of Cape Town which began in 2012 as part of the Mistra Urban Futures program. The situated history and local experience of the ACC has given its researchers unique insights into the challenges and opportunities of co-producing knowledge. These insights show the importance of normativity and point toward the need for iterative reflexivity throughout the research process. The chapter reflects on the experiences of the ACC researchers and how they have navigated the ambiguity of the co-production process.

In Chapter 4, George Mark Onyango and Ben Otieno Obera describe the specific needs for different types of knowledge exchange and co-production in Kisumu, the third largest city in Kenya. This chapter uses the history of the city's development to situate and better understand the complexity of the urban challenges being faced today. Numerous examples are provided of different approaches that have been used by the municipal council to increase collaboration and partnerships with the civil and informal sectors for more effective urban development. The work of the Kisumu Action Team (KAT), an innovative multi-stakeholder forum established in 2003 to address local and regional development needs through collaboration between science, policy, and the private sector, is described in detail. Examples are also given of the work with the Kisumu Millennium Cities Initiative and with the Kisumu Local Interaction Platform of Mistra Urban Futures.

In Chapter 5, Beth Perry and Tim May elaborate upon the rationale behind, and the reasons and preconditions for university–city engagement in Greater Manchester, a northern post-industrial city region in England. This chapter takes a reflexive tour through experiences in working with urban stakeholders to improve the research–practice relationship in the 2000s within the Centre for Sustainable Urban and Regional Futures (SURF) at the University of Salford, Manchester. Grounded in the political–economic context of a UK postindustrial northern city–region, this chapter highlights diverse drivers and motivations for engaging in a knowledge-based urban agenda, the tensions that arise, and the "promise" of co-production in suggesting new approaches to university–city relationships. Perry and May argue that co-production requires a reflexive disposition to guard against co-option and to protect the role of critique.

In Chapter 6, Colin Fudge and John Fien examine some of the Royal Melbourne Institute of Technology's (RMIT) activities regarding the co-production of knowledge for sustainable urban futures in Melbourne and other cities in Southeast Asia. This includes the university's sustainability policy such as its program for green buildings, social entrepreneurship, and indigenous research support, as well as different examples of engaged research on affordable housing, disability day care, climate change adaptation and disaster risk reduction. A number of substantive issues and themes are raised in relation to university–city engagement. These include institutional responsibility and action in varying global settings, learning from the international comparative experience, the role of the researcher in engaging with the complexity of practice, and the management of stakeholders in the co-production of knowledge and practice-embedded research. The potential for knowledge co-production with government, business, and civil society organizations to produce improved urban outcomes and new urban knowledge economies is exemplified and analyzed.

The concluding chapters (Chapters 7 and 8), focus on the implications of different approaches to co-production for their respective practice, policy, and research contexts. In Chapter 7, Tomas Hellström outlines some of the specific quality control issues facing interdisciplinary and transdisciplinary centers, like Mistra Urban Futures, which strive to create socially relevant knowledge through different types of co-production. A number of important challenges, dimensions and functions of evaluation in these types of centers are discussed. Some of these include the need to: bridge the internal and external boundaries of a center, involve stakeholders in mutual learning with researchers, find ways to communicate the complexity of urban problems and co-production, and provide satisfactory descriptions of preliminary impacts. The chapter ends by presenting the outline of a value-based, formative model for evaluating co-production processes at transdisciplinary centers such as Mistra Urban Futures.

In Chapter 8, Merritt Polk returns to opening up a wider discussion on the urban problematic by bringing together the variety of voices and approaches to urban knowledge production in the different city case chapters. The final chapter takes a "meta" view, spanning across the city cases, by comparing and synthesizing the diverse approaches and voices within urban knowledge collected here. The chapter starts by comparing the city cases in relation to the added value from knowledge co-production identified in the cities. It then outlines some of the main barriers and challenges to co-production that were experienced in the different cities. It concludes with a discussion of topics for future research that can support processes for co-producing knowledge for sustainable urban futures.

This book includes a broad and cosmopolitan choice of city cases and examples from intrinsically different urban contexts in Europe, Africa, and Southeast Asia. We target professionals in urban planning and practice as well as researchers and students in different areas of urban-related studies and research. While the majority of the co-authors are researchers, two of the city chapters have been co-authored by practitioners, and all of the co-authors have years of experience working with practitioners from civil society and planning—as consultants, researchers, architects, and planners. These different voices can be distinguished not only by their institutional contexts in practice and research, but also by the different theoretical perspectives and practical experiences they apply to describing and analyzing their cities. We have not limited ourselves here to any one approach to bridging the boundaries within and between practice and research. Instead, while we focus on researcher-initiated co-production, we also provide a continuum of methods, activities, and approaches to co-producing knowledge that include NGO participation in city commissions, seminars between civil servants and researchers, engaged researchers, active intermediaries, and co-led research projects. We hope that this breadth and diversity of examples of knowledge co-production will speak to a wide audience of engaged planners, policy-makers, researchers, and civil society actors in cities around the world.

References

Adams, W. M. (2001) *Green Development 2nd Edition: Environment and Sustainability in the Third World.* London: Routledge.

Albrechts, L. (2012) 'Reframing Strategic Spatial Planning by Using a Coproduction Perspective', *Planning Theory*, 12(1): 46–63.

Batty, M. (2009) 'Cities as Complex Systems: Scaling, Interaction, Networks, Dynamics and Urban Morphologies', in R. A. Meyers, *Encyclopedia of Complexity and Systems Science*. New York: Springer.

——(2012) *Great Planning Disasters: How We Should Tackle Complexity by Taming Wicked Problems.* Downloaded March 17, 2012 from: www.spatialcomplexity.info/files/2013/01/Batty-Hall-Paper-Final.pdf.

Bjurström, A. and Polk, M. (2011) 'Physical and Economic Bias in Climate Change Research: A Scientometric Study of IPCC Third Assessment Report', *Climatic Change*, 108: 1–22.

Boyer, E. (1996) 'The Scholarship of Engagement', *Journal of Public Service and Outreach*, 1(1): 11–20.

Brugmann, J. (2009) *Welcome to the Urban Revolution: How Cities Are Changing the World.* New York: Bloomsbury.

Castro, C. (2004) 'Sustainable Development: Mainstream and Critical Perspectives', *Organization and Environment*, 17(2): 195–225.

Caswell, C. and Shove, E. (2000) 'Postscript to Special Issue on Interactive Social Science', *Science and Public Policy*, 27(3): 220–2.

Davis, M. (2006) *Planet of Slums*. New York: Verso.

Etzkowitz, H. and Leydesdorff, L. (2000) 'The Dynamics of Innovation: From National Systems and "Mode 2" to a Triple Helix of University–Industry–Government Relations', *Research Policy*, 29: 109–23.

Flyvbjerg, B. (2001) *Making Social Science Matter*. Cambridge: Cambridge University Press.

Forester, J. (1989) *Planning in the Face of Power*. Berkeley, CA: University of California Press.

Foucault, M. (1980) 'Truth and Power', in C. Gordon (ed.), *Power/Knowledge: Selected Interviews and Other Writings 1972–1977*. Brighton: Harvester Press, pp. 109–33.

Frank, M. (2002) *The Origins of the Modern World. A Global and Ecological Narrative*. Lanham, MD: Rowman and Littlefield.

Frodeman, R. (ed.) (2010) *The Oxford Handbook of Interdisciplinarity*. Oxford: Oxford University Press.

Funtowicz, S. and Ravetz, J. (1993) 'Science for the Post-Normal Age', *Futures*, 25(7): 739–55.

Gibbons, M., Limoges, C., Nowotny, H., Schwartzman, S., Scott, P. and Trow, M. (1994) *The New Production of Knowledge: The Dynamics of Science and Research in Contemporary Society*. London: Sage.

Godin, B. (2006) 'The Linear Model of Innovation: The Historical Construction of an Analytical Framework', *Science, Technology & Human Values*, 31(6): 639–67.

Griggs, D., Stafford-Smith, M., Gaffney, O., Rockström, J., Öhman, M. C., Shyamsundar, P., Steffen, W., Glaser, G., Kanie, N. and Noble, I. (2013) 'Policy: Sustainable Development Goals for People and Planet', *Nature*, 495(7441): 305–7.

Hall, P. (1988) *Cities of Tomorrow: An Intellectual History of Urban Planning and Design in the Twentieth Century*. Oxford: Blackwell.

Healey, P. (1992) 'Planning through Debate: The Communicative Turn in Planning Theory', *Town Planning Review*, 63(2): 143–62.

——(1997) *Collaborative Planning: Shaping Places in Fragmented Societies*. Hong Kong: Macmillan.

——(2012) 'The Universal and the Contingent: Some Reflections on the Transnational Flow of Planning Ideas and Practices', *Planning Theory*, 11(2): 188–207.

Hessels, L. and van Lente, H. (2008) 'Re-thinking New Knowledge Production: A Literature Review and a Research Agenda', *Research Policy*, 37: 740–60.

Hirsch Hadorn, G., Hoffman-Reim, H., Biber-Klemm, S., Grossenbacher-Mansury, W., Joye, D., Pohl, C., Wiesmann, U. and Zemp, E. (eds) (2008) *Handbook of Transdisciplinary Research*. Berne: Springer.

Holbrook, B. (2010) 'Peer Review', in R. Frodeman (ed.), *The Oxford Handbook of Interdisciplinarity*. Oxford: Oxford University Press, pp. 321–32.

Hudson, B. (1979) 'Comparison of Current Planning Theories: Counterparts and Contradictions', *Journal of the American Planning Association*, 45(4): 387–98.

Hulme, M. (2009) *Why We Disagree about Climate Change: Understanding Controversy, Inaction and Opportunity*. Cambridge: Cambridge University Press.

Huutoniemi, K. (2010) 'Evaluating interdisciplinary research', in R. Frodeman (ed.), *The Oxford Handbook of Interdisciplinarity*. Oxford: Oxford University Press, pp. 309–20.

Huxley, M. (2000) 'The Limits to Communicative Planning', *Journal of Planning Education and Research*, 19(4): 369–77.

Innes, J. E. (1996) 'Planning through Consensus Building', *Journal of the American Planning Association*, 62(4): 460–72.

IPCC (2014) *Working Group II Summary for Policy Makers.* Downloaded May 3, 2014 from: http://ipcc-wg2.gov/AR5/images/uploads/IPCC_WG2AR5_SPM_Ap proved.pdf.

Jackson, M. C. (1991) *Systems Methodology for the Management Sciences.* New York: Plenum Press.

Jacobs, J. (1965) *The Death and Life of Great American Cities.* London: Penguin.

Jasanoff, S. and Wynne, B. (1998) 'Scientific Knowledge and Decision Making', in S. Rayner and E. Malone (eds), *Human Choice & Climate Change*, 4 vols. Columbus, OH: Battelle Press, pp. 1–112.

Kates, R. W., Parris, T. M. and Leiserowitz, A. A. (2005) 'What Is Sustainable Development? Goals, Indicators, Values and Practice', *Environment*, 47(3): 9–21.

Klein, J. T. (1990) *Interdisciplinarity: History, Theory, and Practice.* Detroit, MI: Wayne State University Press.

——(2010) 'A taxonomy of interdisciplinarity', in R. Frodeman (ed.), *The Oxford Handbook of Interdisciplinarity.* Oxford: Oxford University Press, pp. 15–30.

König, A. (ed.) (2013) *Regenerative Sustainable Development of Universities and Cities: The Role of Living Laboratories.* Cheltenham: Edward Elgar.

Lang, D. J., Wiek, A., Bergmann, M., Stauffacher, M., Martens, P., Moll, P., Swilling, M. and Thomas, C. J. (2012) 'Transdisciplinary Research in Sustainability Science: Practice, Principles and Challenges', *Sustainability Science*, 7(supplement 1): 25–43.

Lauria, M. (2000) 'The Limits to Communicative Planning Theory: A Brief Introduction', *Journal of Planning Education and Research*, 19(4): 331–2.

Leydesdorff, L. and Meyer, M. (2006) 'Triple Helix Indicators of Knowledge-based Innovation Systems: Introduction to the Special Edition', *Research Policy*, 35: 1441–9.

Longino, H. (1990) *Science as Social Knowledge: Values and Objectivity in Scientific Inquiry.* Princeton, NJ: Princeton University Press.

Mitlin, D. (2008) 'With and beyond the State – Coproduction as a Route to Political Influence, Power and Transformation for Grassroots Organizations', *Environment and Urbanization*, 20: 339–60.

Nowotny, H., Scott, P. and Gibbons, M. (2001) *Re-thinking Science: Knowledge and the Public in an Age of Uncertainty.* Cambridge: Polity Press.

Oreskes, N. (2004) 'Science and Public Policy: What's Proof Got to Do with It?', *Environmental Science and Policy*, 7(5): 369–83.

Ostrom, E. (1996) 'Crossing the great divide: Coproduction, synergy, and development', *World Development*, 24(6): 1073–87.

Owens, S. and Cowell, R. (2011) *Land and Limits: Interpreting Sustainability in the Planning Process.* 2nd edn. London: Routledge.

Parnell, S. and Robinson, J. (2012) '(Re)theorizing Cities from the Global South: Looking beyond Neoliberalism', *Urban Geography*, 33(4): 593–617.

Perlman, J. (2011) YouTube lecture at EcoCities conference, the mega-cities project.

Pielke, R. (2007) *The Honest Broker: Making Sense of Science in Policy and Politics.* Cambridge: Cambridge University Press.

Pieterse, E. (2011) 'Recasting Urban Sustainability in the South', *Development*, 54(3): 309–16.

Pohl, C., Rist, S., Zimmermann, A., Fry, P., Gurung, G. S., Schneider, F., Speranza, C. I., Kiteme, B., Boillat, S., Serrano, E., Hadorn, G. H. and Wiesmann, U. (2010) 'Researchers' Roles in Knowledge Co-Production: Experience from Sustainability

Research in Kenya, Switzerland, Bolivia and Nepal', *Science and Public Policy*, 37(4): 267–81

Robinson, J. (2004) 'Squaring the Circle? Some Thoughts on the Idea of SD', *Ecological Economics*, 48: 369–84.

——(2008) 'Being Undisciplined: Transgressions and Intersections in Academia and beyond', *Futures*, 40: 70–86.

——(2011) 'Cities in a World of Cities: The Comparative Gesture', *International Journal of Urban and Regional Research*, 35(1): 1–23.

——(2013) 'The Urban Now: Theorizing Cities beyond the New', *European Journal of Cultural Studies*, 16(6): 659–77.

Robinson, J. and Tansey, J. (2006) 'Co-production, Emergent Properties and Strong Interactive Social Research: The Georgia Basin Futures Project', *Science and Public Policy*, 33(2): 151–60.

Rosenhead, J. (ed.) (1989) *Rational Analysis for a Problematic World: Problem Structuring Methods for Complexity, Uncertainty and Conflict*. Chichester: John Wiley.

Russell, A. W., Wickson, F. and Carew, A. L. (2008) 'Transdisciplinarity: Context, Contradictions and Capacity', *Futures*, 40: 460–72.

Rydin, Y. (2007) 'Re-examining the Role of Knowledge within Planning Theory', *Planning Theory*, 6: 52–68.

Sandercock, L. and Forsyth, A. (1992) 'A Gender Agenda: New Directions for Planning Theory', *Journal of the American Planning Association*, 58(1): 49–59.

Sarewitz, D. (2004) 'How Science Makes Environmental Controversies Worse', *Environmental Science and Policy*, 7(5): 385–403.

Satterthwaite, D. (2005) *The Scale of Urban Change Worldwide 1950–2000 and Its Underpinnings*. Human Settlements Discussion Paper Series, Theme: Urban Change – 1. IIED.

Saxe, J. G. (1872) *The Blind Men and the Elephant*. Downloaded December 2, 2013 from: http://en.wikisource.org/wiki/The_poems_of_John_Godfrey_Saxe/The_Blind_Men_and_the_Elephant.

Talwar, S., Wiek, A. and Robinson, J. (2011) 'User Engagement in Sustainability Research', *Science and Public Policy*, 38(5): 379–90.

Trencher, G., Yarime, M. and Kharrazi, A. (2013) 'Co-creating Sustainability: Cross-sector University Collaborations for Driving Sustainable Urban Transformations', *Journal of Cleaner Production*, 50(1): 40–55.

Turnhout, E., Stuiver, M., Klostermann, J., Harms, B. and Leeuwis, C. (2013) 'New Roles of Science in Society: Different Repertoires of Knowledge Brokering', *Science and Public Policy*, 40: 354–65.

Turok, I. and Parnell, S. (2009) 'Reshaping Cities, Rebuilding Nations: The Role of National Urban Policies', *Urban Forum*, 20: 157–74.

UN (2013) *The Millennium Development Goals Report 2013*. United Nations. Downloaded July 13, 2013 from: www.un.org/millenniumgoals/pdf/report-2013/mdg-report-2013-english.pdf.

UN DESA (2009) *2009 Revision of World Urbanization Prospects*. Downloaded May 8, 2011 from: http://esa.un.org/unpd/wup/index.htm.

——(2010) *UN Press Release: World Population to Reach 10 Billion by 2100 If Fertility in All Countries Converges to Replacement Level*. Downloaded May 8, 2011 from: http://esa.un.org/unpd/wpp/Other-Information/Press_Release_WPP2010.pdf.

UN-HABITAT (2010) *State of the World's Cities 2010/2011: Bridging the Urban Divide*. Nairobi: UN-HABITAT.

United Nations (1992) *Agenda 21: United Nations Conference on Environment and Development, Rio de Janeiro, Brazil.* Downloaded July 13, 2013 from: http://sustainabledevelopment.un.org/content/documents/Agenda21.pdf.

——(2012) *World Urbanization Prospects: The 2012 Revision.* UN Department of Economic and Social Affairs, Population Division. Downloaded July 13, 2013 from: esa.un.org/unpd/wup/index.htm.

Venditti, M., Reale, E. and Leydesdorff, L. (2013) 'The Disclosure of University Research to Third Parties: A Non-market Perspective on an Italian University', *Science and Public Policy*, 40(6): 792–800.

Watson, V. (2003) 'Conflicting Rationalities: Implications for Planning Theory and Ethics', *Planning Theory and Practice*, 4(4): 395–407.

——(2014) 'Co-production and Collaboration in Planning – The Difference', *Planning Theory & Practice*, 15(1): 62–76.

Weingart, P. (2008) 'How Robust Is Socially Robust Knowledge?', in M. Carrier, D. Howard, and J. A. Kourany (eds), *The Challenge of the Social and the Pressure of Practice: Science and Values Revisited.* Pittsburgh, PA: University of Pittsburgh Press.

——(2010) 'A Short History of Knowledge Formation', in R. Frodeman (ed.), *The Oxford Handbook of Interdisciplinarity.* Oxford: Oxford University Press, pp. 1–14.

Wiek, A., Ness, B., Schweizer-Ries, P., Band, F. S. and Farioli, F. (2012) 'From Complex Systems Analysis to Transformational Change: A Comparative Appraisal of Sustainability Science Projects', *Sustainability Science*, 7(supplement 1): 5–24.

Wiek, A., Talwar, S., O'Shea, M. and Robinson, J. (2014) 'Toward a Methodological Scheme for Capturing Societal Effects of Participatory Sustainability Research', *Research Evaluation*, 23: 1–16.

Wikman, A. and Rockström, J. (2011) *Den stora förnekelse.* Stockholm: Medströms Bokförlag.

World Commission on Environment and Development (WCED) (1987) *Our Common Future.* Oxford: Oxford University Press.

Zierhofer, W. and Burger, P. (2001) 'Disentangling Transdisciplinarity: An Analysis of Knowledge Integration in Problem-Oriented Research', *Science Studies*, 20(1): 51–74.

2 Co-producing knowledge in Gothenburg through political collaboration, cross-sector programs, triple helix science parks, and Mistra Urban Futures

Björn Malbert, Merritt Polk, and Stig Egnell

Introduction

In 2010 Mistra Urban Futures, a transdisciplinary center for promoting sustainable urban development was established in Gothenburg in response to a growing understanding that different types of knowledge co-production were needed to better address the current and future challenges facing the region. This insight stems in part from the history of the city, where external threats and internal crises have fostered a culture of collaboration, sometimes referred to as the "Gothenburg Spirit" (Falkenmark 2010). In the early twenty-first century, the political parties governing Gothenburg initiated a collaborative governance approach, which was based on earlier experiences of cooperation between the city and local business communities, including such major industries as the City shipyard (Götaverken), the Swedish ball-bearing company (SKF), Volvo, and Ericsson, and on voluntary collaborations between municipalities in the greater Gothenburg area around issues relating to sustainable growth, public transportation, and social segregation (Elias *et al.* 2011; Montin *et al.* 2014; Polk 2011). These efforts culminated in numerous in-depth partnerships between academia and private and public sector actors.

This chapter begins with a brief overview of the historical background of modern Gothenburg. The subsequent sections present the main internal and external driving forces of knowledge co-production. This is followed by concrete examples from the public sector and academia, including the joint effort to set up and run Mistra Urban Futures. The chapter concludes with lessons learned from the many co-production projects and activities that have been underway in Gothenburg since 2010.

Gothenburg's history in brief

The drivers of co-production in Gothenburg stretch back to the city's founding in 1621, on a battleground between Denmark, Sweden, and Norway. Gothenburg was founded at the mouth of the Göta River, Sweden's only point of access to the North Sea, which at the time was flanked by Norway and

Denmark to the north and south, respectively. The well-preserved fortresses at Marstrand, Kungälv, and Älvsborg bear witness to the strategic importance of the site.

Shipping and trade were the early drivers of Gothenburg's development

In the following centuries, Gothenburg was the main Swedish port to the North Sea; shipping and trade were thus crucial components in the early development of the town. Not least, the East Indian Company (1731–1813) had a strong influence on Gothenburg's urban development at the time. Having made huge profits from trade with China and other countries, the company's owners invested in buildings and founded institutions many of which remain to this day, Sahlgrenska University Hospital and Chalmers University of Technology being two prominent examples (Falkenmark 2010). The early physical and administrative improvements of the town were designed and carried out by experts brought in from other European countries, especially Holland (Peterson 2010). The result was a town surrounded and criss-crossed by canals. Some of these are still visible in the city center of today; others have been filled in, and in some cases, turned into tram tracks. During this period, the executive power in the city was in the hands of an elite group of rich shipowners, traders, and merchants within the frames of a town charter, which was granted and controlled by the state (Falkenmark 2010).

Prior to the twentieth century, Gothenburg was a rather small harbor and trade city (20,000 inhabitants in 1850 and about 130,000 in 1900, compared with about 520,000 in 2012) (Hansson 1999; SCB 2014; Warburg 1913). The emerging handicrafts and industries developed outside the formal borders of the city along with informal settlements for the growing labor population. Different transformative movements and struggles characterized the decades before and after the turn of the twentieth century (Thörn 2010). Technical innovation and industrial production became the new drivers for the development of Gothenburg. The Swedish ball-bearing manufacturer (SKF) and Volvo are examples of present-day companies that were founded and developed during this period. SKF had its main factories in the Old Town district (*Gamlestaden*) northeast of today's city center, where the city originally was established. The car production at Volvo started through initiatives from a group of engineers from SKF. Volvo grew to be, and still is, a very important industrial company in Gothenburg. Its factories, located on Hisingen, a large island on the north side of the Göta River, have expanded over the years. With its large-scale manufacturing and logistics, Volvo has had a great impact on the planning and development of the Gothenburg metropolitan area. Furthermore, shipping and trade were also very significant, as manifested in growing harbor activities, as well as an expanding shipyard and wharf industry located in central parts of the city at Götaverken, Lindholmen, and Eriksberg.

The city expanded its administrative borders continuously by incorporating surrounding municipalities up to the early 1970s starting with Lundby on Hisingen in 1906 (Andersson 1993). New players appeared in the political arena, including industrialists and capital owners, on one side, and the labor force and trade unions on the other (Thörn 2010). The glaring need that arose for improved living conditions, sanitation, education, health care, and communications was addressed, sometimes through small-scale private initiatives. Later on, such initiatives were taken over and up-scaled by the public sector, especially during the long Social Democratic era and the efforts to make homes for all. "The People's Home" (*folkhemmet*) became the slogan for shaping a modern Swedish society with tax-funded welfare services available to all inhabitants.

The Social Democratic Party, founded in 1889, attained a majority in the Gothenburg Municipal Council for the first time in 1922. Even in victory, the new political leaders decided to keep some of the more experienced politicians from the defeated conservative minority in key positions for the political management of the city. This is said to have been the start of what later came to be known as the "Gothenburg spirit," meaning a specific capacity in Gothenburg for different parties and interests to cooperate on challenging development issues. Other references to the "Gothenburg spirit" can be found in both earlier and later urban development processes (Falkenmark 2010).

The first half of the twentieth century saw strong growth within both the private and public sectors in Sweden, despite the interruptions of two world wars and the 1930s Depression. This growth culminated in the late 1960s and early 1970s. For the public sector, the post-war period from late 1940s forward was especially important. The political vision of the "People's Home" included investments in all levels of education. It also included the design and implementation of public sector institutions. These two trends together became the starting point for the building of the Swedish welfare state during forty-four years of continuous Social Democratic rule. The higher education system became finely tuned to the needs of the welfare state, with experts from different professional education sectors being recruited to fill positions in the steadily expanding public sector (Thörn 2010). In 1959, the new comprehensive plan for Gothenburg manifested the visions of the modern industrial city. Decent housing and private cars for all households were also incorporated as important social welfare goals (GPL 1959). Zoning was the preferred technique of city planning at this time, with the different functions of the city being separate yet linked together by large-scale systems of roads and tram lines.

Economic crisis and stagnation—the second half of the twentieth century

The 1960s were the peak decade for the growth of Gothenburg as a heavy industrial city. The expanding industry required more workers who in turn required housing. Labor immigrants from the countryside, as well as from countries like Finland, Greece, and Italy, came to Gothenburg on their own initiatives, or as a result of active recruitment by industrial companies (Thörn 2010). The "Million Apartments Program" (*Miljonprogrammet*), aiming to build one million new apartments within a ten-year period, was launched to deal with the acute lack of housing (Soidre 2010). These apartments were located on the outskirts of town as a reaction against the unhealthy overcrowding in the city center and still have an important structural impact on the city.

In the 1970s, Gothenburg's shipbuilding industry collapsed. Over the course of a few years it went from being world leading to almost nothing (Poulsen and Sornn-Friese 2011). There were two major effects of the crisis. First, reductions in the industrial labor force resulted in empty apartments in the Million Apartments Program areas, creating an economic burden for the municipal housing owners. Second, the crisis gave rise to a number of radical opposition movements, especially among students—the so-called 1968 generation—including an environmental movement often referred to as the "green wave" of the 1970s. The latter received political recognition at the national level, as manifested in Stockholm's hosting in 1972 of the United Nations Conference on the Human Environment, the first global summit of its kind.

During the 1980s, when the housing industry was weak, the construction industry focused on commercial buildings. This led to a boom and a subsequent collapse of the real estate market in Gothenburg (Jörnmark 2005). At the same time the city became the owner of vast properties previously occupied by wharf and harbor activities along the Göta River. In order to manage the stagnation in urban development, the Municipal Council established a new municipally owned company, Älvstranden Utveckling AB, in 1996. It was owned and guided by the Municipal Council, but had the capacity to act according to the rules of the marketplace. This ownership structure made it possible to undertake an ambitious strategy of coordinated planning, that is, to encourage close cooperation between the public and private sectors in Gothenburg. The challenge was to find a way to utilize this large, centrally located tract of land to transform Gothenburg from a heavy industrial city into a more diverse city oriented around culture, knowledge, and technology. In tangible terms, the goal was to develop a mixed commercial and residential area with good environmental quality. The "Gothenburg spirit" showed itself with the forming of a close collaboration between the city, the academic community, and the business sector, in what has been called a "triple helix" approach (Etzkowitz and Leydesdorff 2000).

Towards sustainable urban development—the end of the twentieth century and the first decade of the twenty-first

A number of environmental projects were launched by the then Swedish minister for the environment at the end of the 1980s (Lundqvist 2000). These projects were state funded and aimed to address severe environmental problems in Gothenburg and other selected parts of the country. The conditions for receiving the funding were not so much about what the project should work on and how, but rather who should be involved. The ambition was to engage and involve experts outside the formal agencies. In that sense the Gothenburg Environmental Project can be regarded as an early attempt to promote collaboration between actors with different backgrounds and positions. For instance, the political chairperson was already well known before the project for her strong interest in social issues. She later became the leader of a task force under the Municipal Council that focused on environmental issues. The project leader, on the other hand, was recruited from the Greenpeace movement. The Gothenburg Business Region, a company tasked with promoting economic growth in the Gothenburg region, later hired him. The project supported yet another side of the so-called "Gothenburg spirit" by creating new kinds of collaborative networks in the city (Wärmby 1992). In these networks quite a few of the environmental activists of the 1970s now occupied formal positions in important NGOs as well as private and public sector agencies.

At the 1992 World Summit in Rio de Janeiro, the concept of sustainable development, as outlined by the Brundtland Commission (1987), was made operational in the Agenda 21 declaration and other documents. This led many Swedish municipalities, not least Gothenburg, to develop strategies and take concrete measures to promote sustainable development. Gothenburg adopted what was at the time a somewhat radical environmental policy. The Comprehensive Plan of 1993 aimed to integrate the challenges of environmental carrying capacity and economic competitiveness (ÖP 1993). According to the plan, Gothenburg should be developed through measures in which these two basic goals overlap. This approach made it possible to get strong support from all political parties on the Municipal Council. The plan was presented at the Habitat II Conference in Istanbul (1996) as an example of best practice.

The theme of sustainable development was followed up in the next Comprehensive Plan (1999) where the necessity of citizen involvement was captured in the concept of citizens' power (ÖP 1999). This indicated a stronger political interest in the social aspects of sustainable development connected to emerging social problems in the city. One reason can be traced to the many empty apartments in the suburban areas of Gothenburg after the structural changes of the 1970s labor market. These now became a resource when Gothenburg, in line with national immigration policies, opened its doors to refugees from all around the world. The result is a number of housing areas built during the Million Apartments Program that are dominated by first- and second-generation immigrants and refugees from various regions of the world who have suffered and still are suffering from the warfare and social instability of the last five decades. These neighborhoods are among the most highly segregated housing areas in all of Europe (Holgersson *et al.* 2010).

At the same time, many young Swedish families have moved away from Gothenburg to neighboring municipalities where they can find apartments or houses of high standard at lower costs than in Gothenburg. This has contributed to urban sprawl, which in turn requires major investments in infrastructure for commuting. Most of the region's workplaces are still located in the city proper. Taken together, these two trends have resulted in a segregated urban region where various kinds of tensions between people with different living conditions pose serious challenges for urban development. The first decade of the twenty-first century was characterized by several attempts to respond to these and other challenges facing the urban region. Some of these will be discussed below.

As this brief history shows, over time Gothenburg has gone through various crises not very different from those experienced by other industrial and harbor cities in Europe due to economic globalization and the structural changes brought about for manufacturing industries, as well as immigration trends. However, in Gothenburg many local responses to these crises, and to emerging challenges related to climate change, resource constraints, and social polarization, have been based on collaboration between stakeholders and actors from different spheres of society. This has further been enhanced by a specific culture of collaboration, the "Gothenburg Spirit," through which the insight into the need for co-production of knowledge and solutions has matured. However, this culture of collaboration has also been severely criticized for being elitist and run in a top-down fashion, with weak involvement of local communities and citizens' interest groups. The result has been collaboration between the city, the business sector, and the universities to promote investments in a culture-, tourism-, and knowledge-based city. The aim is to attract a well-educated middle class to Gothenburg, but, in practice, it has also strengthened the growing social polarization of the city (Thörn 2010).

Drivers of co-production and responses at the national level in Sweden

Predicted and ongoing climate change, resource constraints, and structural changes in the economy, industrial production, and labor markets, as well as the increasing streams of refugees from wars and conflicts, have become drivers of policy-making at the national as well as local level in Sweden. Environmental concerns, among others, were expressed in sixteen ambitious national environmental goals (Swedish Institute 2013). Other major issues concerned the growth of and competition between the three large urban regions in southern Sweden: Stockholm, Gothenburg, and Malmö. This overall process of urbanization has led to stagnation and depopulation in other parts of the country, where an aging population requires more social services than can normally be afforded by a shrinking tax base.

Within the three growing urban regions, migration from cities can also be observed. This has been caused by high costs of living and lack of housing in town centers. Two main challenges accompany this suburban sprawl. One is related to the expanding housing and labor markets of the urban regions, which in turn require huge investments in infrastructure to transport people and goods without increasing the already existing dependence on private cars. The second is the problem of social segregation mentioned above, whereby urban sprawl spatially separates people with different economic, ethnic, and social backgrounds. In all three main cities, immigrants from all over the world are concentrated in less attractive urban districts, in the suburbs of Stockholm and Gothenburg, or the more central areas of Malmö. In all three cities, these concentrations mainly occur in the apartment blocks from the Million Apartments Program. The regional and local responses to these challenges that have led to the co-production of knowledge and solutions in the Gothenburg region will be discussed later in this chapter.

The concept of sustainable development has been implemented at all levels of government in Sweden, though with differing interpretation and emphasis over time. After the Brundtland Commission report in 1987, national policy documents were formulated to promote ecologically sustainable development, based on more than one hundred years of strong traditions and institutions for environmental protection in Sweden. This put the concept in the hands of formal and informal environmental organizations and interest groups that pushed the issues in greater or lesser conflict with other social and economic interests.

After the 1992 World Summit in Rio and the launch of Agenda 21, a more holistic view of sustainable development was introduced based on the three pillars of social, economic, and environmental concern. Many Swedish municipalities appointed so-called Agenda 21 coordinators, and efforts to implement the three-dimensional concept of sustainable development were often linked to comprehensive spatial planning, as in the example from Gothenburg mentioned above (Khakee 2010). However, in the first decade of the twenty-first century a new bias became apparent in the notion of sustainable growth, where economic growth became the main goal of sustainable development applied in the region. This is based on the belief that it is only through economic growth that environmental and social challenges can be addressed (Cullberg *et al.* 2014; Polk 2010).

National research policy promoting the co-production of knowledge

On the national level, the complexity of the issues related to sustainable development led to a growing insight that no existing social sector or scientific discipline could claim to have the knowledge and capacity to solve existing social and environmental problems on their own. This insight had an impact on national research policy. National research funding organizations were reorganized around the turn of the millennium, and policy-directed calls for research applications became more frequent. These often required cross-disciplinary approaches and external stakeholder involvement. While the Swedish Scientific Research Council is supposed to continue to be responsible for basic research, the newly established research council for the environment, forestry, agriculture, and the built environment, FORMAS, is meant to be responsible for the named sectors as well as for research that supports sustainable development (see respective web pages). Another research council, Vinnova, was set up to promote the innovation of products and services in support of Swedish production and export (ibid.).

Another important actor at the national level is the Swedish Foundation for Strategic Environmental Research, Mistra (see web pages). Mistra is responsible for funding research of high scientific quality that has both a direct impact on practice and is supportive of Swedish exports of products and services. In the last few years research on urban development issues has been recognized and supported by different types of funding from these national organizations and, as already mentioned, has often employed cross-disciplinary approaches and broad stakeholder involvement. This, in turn, has influenced universities' research policies and organization. Some examples of how the universities in Gothenburg have responded to these challenges through different types of collaboration with non-academic actors will be outlined below.

Examples of responses from policy and planning in the region

In the Gothenburg area, the co-production of knowledge began with different types of practice-oriented collaborations and networks. The examples of multilevel governance and cross-sector collaboration described here are a response to the major political and administrative restructuring that took place in western Sweden in 1999 and the early 2000s. A number of regional bodies, primarily responsible for health care issues, were merged into the Västra Götaland Region (VGR), which serves around 1,500,000 inhabitants. VGR is divided into four sub-regions, or associations of local authorities, of which the Gothenburg Region Association of Local Authorities (GR) is the largest and most influential. VGR is a political body with an elected representative council and its own tax base. Apart from health care institutions, VGR also manages environmental, cultural, and regional development issues, some of which were taken over from the national county administration. GR is a voluntary political collaboration between the thirteen municipalities surrounding Gothenburg. The regional arm of the national government, the County Administration Board (CAB), was also restructured. The CAB in Västra Götaland resulted from the merging of three smaller county administrations. These major political and administrative changes have had a great impact on many of the examples of knowledge co-production described here. The establishment of new organizations and institutions often makes possible (as well as requires) creativity and innovation in finding the proper forms and processes for action.

VGR's vision of the good life

The first example of state-initiated co-production is the development of a vision for "The Good Life" in Västra Götaland (VGR 2005). The new regional body had to formulate a vision to guide future development, well aware that this vision could not be implemented by decisions in the regional council, but only through the active participation of the forty-nine municipalities in the region, the private sector, and the inhabitants of the region. The vision was accordingly designed together with politicians, regional officials, external experts, and selected representatives of academia in thematic groups addressing environmental, social, and economic aspects of sustainable development. At the time, it was an ambitious vision, aiming to integrate the three dimensions of sustainability and, for the first time in this type of document, emphasizing their equal importance. It has subsequently been adapted to local conditions and demands in the different sub-regions. The GR version of the vision was titled *Sustainable Growth*, reflecting the shift of emphasis within the concept of sustainable development mentioned above (GR 2006).

The political consultation process in the Gothenburg region 2002–2008

The Gothenburg Region Association of Local Authorities (GR) has the legal right to develop regional development plans and guidelines for municipal physical planning. However, when the work on a new regional plan was to be started in early 2000, leading politicians and civil servants at GR wanted to try another approach. A new type of voluntary consultation and deliberation process was therefore introduced between and among the politicians in the thirteen municipalities of the region (GR 2005). GR does not have the power to implement regional policies and strategies; this is reserved for the associated municipalities. Thus, hosting direct consultations with politicians about current and future challenges to the region and possible ways to address them was regarded as more relevant than drafting traditional planning documents that risk collecting dust on a shelf. Since 2002, five such rounds of consultation have been held resulting among other things in two strategic documents for the region: *Sustainable Growth – Goals and Strategies with a Focus on Sustainable Regional Structures* and the *Structural Illustration for the Gothenburg Region* (GR 2006, 2008). In an undated brochure, GR describes the process as: "a broad, unbiased and constructive dialogue around important regional issues. Through the consultation process, mutual understanding is created of the importance of regional co-operation, including having a shared attitude towards planning matters in general and a common view of instruments and working methods" (GR 2008). In practice, this process has involved five rounds of consultations consisting of three-hour dialogue sessions in each of the thirteen municipalities, with approximately 1,100 local politicians taking part in each round. The politicians have been very appreciative of these consultations, and of how the needs and views of the neighboring municipalities have been taken into consideration in decision-making for the City of Gothenburg (Elias *et al.* 2011; Cullberg *et al.* 2014).

HUR2050, sustainable development for the region

In parallel to the political process mentioned above, the Swedish Road Administration, Region West, started the HUR2050 network in 2002 to create an arena for dialogue and collaboration to better address the complexity and long-term nature of problems and solutions for sustainable transportation and urban development in the Gothenburg region. The network was composed of around thirty-five directors and managers from fourteen national, regional, and local political, regulatory, and operating agencies within the areas of transport and urban planning. Together these agencies covered a diversity of jurisdictions and mandates, including road and rail planning, transit operations, regional development, urban planning, and environmental protection. Although HUR2050 was an informal network and was not directed or supervised by any higher authority, it was unique in being sanctioned by all of the participating agencies and, furthermore, in having regular contact with leading local and metropolitan politicians. Overall, HUR2050 contributed greatly to establishing a working network of transport and city planners who together explored how long-term sustainability based on equal consideration of environmental, economic, and social aspects could be implemented in the region (Polk 2010, 2011).

K2020, sustainable public transport in the Gothenburg region

K2020 was started in 2003, as a working group of HUR2050, with the aim of analyzing the future of public transport in the city region. It was composed of six organizations which were responsible for different aspects of the planning and operation of public transport in and around Gothenburg: the Gothenburg City Transport office, the Gothenburg Region Association of Local Authorities (GR), Västra Götaland Region, Västtrafik, the Swedish Road Administration, and the Swedish Railroad Company in the Western Region. The work with K2020 began with an analysis of the current transport situation in the region, and continued with various thematic working groups that studied alternative development opportunities for public transport in terms of infrastructure, transport modes, and future travel needs. A report was submitted in the spring of 2005 in the form of a preliminary scorecard for public transport in the Gothenburg region in 2020. This formed the basis for the decision to reorganize K2020 as a political project with the board of GR as the steering committee. The management team consisted of senior staff from each participating organization. A number of reports documenting the work are available for download on GR's website (see the K2020 project website). The total budget for the project was about SEK 10 million (approx. 1.4 million US dollars, or 1 million euros), which was provided by the organizations involved. Joint ownership through co-financing was a deliberate strategy to create a sense of shared responsibility for project implementation.

In 2009, a consensus decision was taken in GR to adopt the K2020 vision, which was to increase the share of public transport from approximately 28 percent at the time to 40 percent by 2025. To reach this vision, the region's entire projected increase in passenger journeys would have to be absorbed by public transport. The visible results of K2020 are these consensus decisions and the final report, which all of the municipal councils in GR approved. Several of the surrounding municipalities also conducted local K2020 projects to analyze possible ways to translate the regional vision into local terms. In addition, different expert groups have also produced a number of reports in the project. The GR decision caused the politicians to raise their target for the share of trips by public transport from the official proposal of 35 percent to 40 percent with an extension of the completion date to 2025. This is an ambitious goal given the current capacity constraints in and around Gothenburg.

K2020 has made significant contributions to the discourse around sustainable transportation in the form of new knowledge and attitudes on individual and organizational levels as well as in the region as a whole. Civil servants speak of a *paradigm shift*, where the central importance of public transport for the region's growth has replaced the previous central role of the automobile (Polk and Malbert 2010). This is especially remarkable in a region long dominated by the automobile industry and where earlier urban development has largely promoted car use. There is a strong understanding that continued collaboration, consensus, and coordination are both necessary and possible in the future development of the city and region. K2020 has become something of a symbol of a holistic approach to urban planning by integrating knowledge and expertise from different sectors and levels of governance and bridging the gap between civil servants and politicians.

S2020, and the search for the social aspects of sustainability

Segregation is one of the most complex and persistent problems in the Gothenburg region. As we have shown in the brief history above, the seriousness of segregation in Gothenburg is due to a combination of the location of available housing with immigration policies and regional employment trends. Social sustainability in urban development is a priority for all departments and public companies in the region. However, despite the large amount of political attention directed to reducing segregation and the differences in income, employment, health, and education that segregation entails, the situation continued to worsen. In response to this, in 2007, the political leadership of Gothenburg created a task force at the Department for Social Resources. This task force was modeled after the success of K2020, and was called S2020 (Cullberg *et al.* 2014).

The initial task of S2020 included identifying experts and researchers who could support the city in its fight against segregation by giving social issues the same status and consideration as economic and environmental issues. Today, S2020 serves a supporting role in the city's work with social sustainability. The team cooperates with all city departments in Gothenburg on developing capacity, influencing opinions, building networks, and initiating research on different aspects of social sustainability. Its current vision for the city focuses on critical issues such as the distribution of prosperity and the ability of citizens to influence decisions through different types of participation, ownership, and access to the city. Achieving joint responsibility for social sustainability is seen as a key challenge for all authorities in the city.

Overall, S2020 supports politicians and officials of the city with knowledge and methods for citizen dialogue. The S2020 website has become a new platform and meeting place for anyone who is interested in the development of Gothenburg in a socially sustainable direction. It is a meeting place for professionals, researchers, students, and interested private persons, and serves as a network for discussions and cooperation. The platform also includes a number of models for supporting work on social sustainability. One example includes the "Knowledge Matrix," a digital knowledge bank where visitors can study research on social environments, look at examples from districts in Gothenburg, or add their own examples. Different themes such as "a holistic city," "well functioning everyday life," and "safety and openness" are coupled to different geographical levels such as a single building, an urban district, the city as a whole, or the entire region. The Knowledge Matrix has been developed as a package using different analytical tools including "Social Consequence Analysis" and "Children's Consequence Analysis". These tools are the result of extensive collaboration between researchers and practitioners. It is hoped that the exchange of ideas, experiences, and information will serve as a source of inspiration and input in people's daily work with these issues. This S2020-created platform represents a deliberate effort by the City to initiate and support solutions to high-priority issues and challenges facing the Municipal Council today.

Examples of responses from the academic community in Gothenburg

The relatively new focus of Swedish national research policy on cross-disciplinary research approaches and stakeholder involvement, together with internal discussions on the future role of the universities and their relations to the surrounding society, have resulted in changes in research and education strategies. There has always been interaction between researchers from different disciplines as well as between researchers and practitioners, but the success and continuity of such work has depended heavily on individual initiatives, capacities, and relations. The third mission of the university, after education and research, to contribute to society outside the academy has often been mentioned, but has been less actively promoted and supported by university leadership. The following examples of initiatives to support different types of knowledge exchange and co-production are just a few of many responses that can be found at Chalmers University of Technology (Chalmers) and the University of Gothenburg (GU).

The concept of Science Parks

Lindholmen Science Park, is one of the new platforms for collaboration between the universities, the city, and other partners with shared interests. The idea is to support innovations in fields of common interest through an open source approach. This is the opposite of the approach that previously was dominant in Technology Parks, where selected researchers worked with people from a specific company in closed rooms. Instead, the idea is to invite competing companies to work together with relevant researchers to solve common problems and share knowledge. The City of Gothenburg, the universities, and companies from the private sector have established three science parks, within IT, logistics, and transport (Lindholmen Science Park), and one on materials and the built environment (Johannesburg Science Park), and within the life sciences (Sahlgrenska Science Park).

The knowledge triangle, knowledge clusters, and areas of advance

Chalmers has an overall vision that research and education should contribute to "a sustainable future" (2010). This vision emphasizes the importance of integrating sustainable development and an understanding of the interplay between people, technology, and society, into all educational programs. In the first decade of the twenty-first century, internal discussions on university–society relations resulted in approaches conceptualized as the knowledge triangle, knowledge clusters, and areas of advance. The knowledge triangle stands for the integration of research, education, and innovation. At Chalmers, the university is seen as a driver of the knowledge-based society through the knowledge triangle, where the university provides the necessary capacities and competencies for sustainable development. According to this approach, the concept of the knowledge triangle should be put into practice in thematic knowledge clusters that involve the public and private sectors.

Taking such an active role in society requires rethinking the internal organization and reward systems of the university. In January 2007, three new vice presidents were appointed at Chalmers. Their mission was to implement the vision of "Chalmers for a sustainable future" by increasing collaboration and reducing barriers within the university itself as well as between Chalmers and external stakeholders. This resulted, among other things, in the establishment of eight Areas of Advance: the Built Environment, Energy, Information and Communication Technology, Life Science Engineering, Nanoscience and Nanotechnology, Materials Science, Production, and Transport. Each Area of Advance provides neutral meeting places where research, education, and innovation at Chalmers can be internally linked and made visible to the surrounding world. Individual faculty members are still members of their departments, however they are also active and visible in one or more of these areas. Given that universities are often more stable than industries and contain all three components of the knowledge triangle—research, education and innovation—they can take on a unique role in building such knowledge clusters in neutral, open and inviting ways (Holmberg 2012).

Examples of high-level collaboration between academic and political organizations in the region

The above-mentioned responses within practice and academia respectively have paved the way for more frequent and intense cooperation between the university and the city–region. In addition to the areas of advance described above, the political and academic leaders in the region have joined together to discuss the issues that they see as important for the future development of the region.

The first example of high-level collaborative work between academic and political organizations is the Dialogue Group, established in 2009. This group brought together the top management of five central organizations in the region in a series of meetings. The organizations include Chalmers, GU, the City of Gothenburg, VGR, and the West Sweden Chamber of Commerce, a private organization promoting the interests of the business community in the region. In 2011, the leaders of these five organizations wrote a debate article in the local newspaper that positioned western Sweden as a "test arena for the future" (Andersson *et al.* 2011). This article also presents a number of factors that are vital to creating a sustainable region. These include:

- Building trust within the organizations, so that employees dare to think creatively and take risks.
- Ensuring low internal organizational barriers, to welcome and promote commitment and collaboration within and across organizations.
- Supporting cooperation and learning between organizations, so they can become each other's ambassadors.
- Promoting and maintaining openness to the outside world.
- Acknowledging that not only sustainable development but also people's needs are the driving forces of regional growth.

(Polk *et al.* 2013: 175–6, translated from Anderson *et al.* 2011)

As all of these factors aim to promote openness and collaboration between the participating organizations, they amount to a continuation of the collaborative tradition in the region described earlier in this chapter. The Dialogue Group also identified a number of knowledge clusters for the region. These include: Life Science, Urban Futures, Bio-based Products, Mobility Solutions, and Marine and Maritime. Each of the five organizations has taken responsibility for mapping one of these areas and developing concrete forms of collaboration within it (Five Clusters 2012).

Joint responses from practice and research: creating partnerships for knowledge co-production at Mistra Urban Futures

The second example of high-level collaboration between academic and political organizations is Mistra Urban Futures. As noted in the introduction to this chapter, Mistra Urban Futures can be seen as the culmination of a number of trends for in-depth collaboration and knowledge exchange in the Gothenburg region. The history of Gothenburg explains some of the reasons why it was possible to initiate and successfully operationalize such a center in the region. This is the backdrop against which the involved public and academic organizations initiated the partnership that was to become Mistra Urban Futures. In this section, we will describe how this was achieved in practice and give examples of some of the strategies and activities that form the foundation of the Center. While Mistra Urban Futures importantly also includes four international partners in three different global regions, this section will focus on the Center Headquarters and activities in Gothenburg. The other platforms and research partnerships and the work with different forms of co-production in Cape Town, Kisumu, Greater Manchester, and Melbourne are described in Chapters 3–6.

Mistra Urban Futures—origins, organization, and strategies for promoting co-production

Mistra, the Swedish Foundation for Strategic Environmental Research, initiated and is a major investor in Mistra Urban Futures, as is Sida, the Swedish International Development Cooperation Agency. The initial call for bids by Mistra to set up a transdisciplinary center stipulated the requirements of matching funds and public–university partnerships. These requirements fell upon fertile ground in Gothenburg and were taken as an opportunity to build upon the different examples of political- and practitioner-based cross-sector and multilevel collaboration that the municipal and regional organizations had been working with since the 1990s. However, given the common experience that research organizations initiate and maintain control of research applications, it was crucial, from the very beginning of the process, to create inclusive processes that required mutual commitment and responsibility from both the public and research organizations. A great deal of effort has been put into ensuring that the Center not only "talks the talk," but also "walks the walk" of a transdisciplinary center based upon knowledge co-production. The participatory process of writing a competitive, and later successful application bid required clear strategies to create a high degree of commitment and joint responsibility at the Center. The result is a unique construction of researcher-initiated co-production, where research and practice meet on equal terms to co-produce knowledge for sustainable urban development (Polk *et al.* 2009).

One of the most important characteristics of Mistra Urban Futures is the organization of the Center itself. This is based on co-funding, shared ownership, and joint leadership from public, academic, and research funding organizations. The Center is made up of the Gothenburg Consortium, which includes the four governing and regulatory levels in Sweden, as well as three research organizations. Among the public organizations, the City of

Gothenburg and the Gothenburg Region Association of Local Authorities (GR) represent the local and metropolitan levels. Representatives of the regional level include Västra Götaland Region (VGR), the provincial political body, and the County Administration Board (CAB), the regional arm of the national government. The research and educational organizations include Chalmers, the host institution, the University of Gothenburg, and IVL, the Swedish Environmental Research Institute. These seven partners together combine all of the political, administrative, and regulatory public agencies with the main educational and research institutions in the region. In addition to the local platform in Gothenburg, the Center also has three affiliated platforms in Cape Town, Kisumu, and Greater Manchester, and a number of strategic partnerships with local, regional, national, and international actors.

The first strategy for creating commitment and joint responsibility within the Gothenburg Consortium partners was financial in nature. Co-financing was reformulated from being a requirement for Mistra funding, to being a central strategy for promoting joint responsibility and commitment to the Center. In practical terms, the requirement for co-financing in times of limited economic resources meant that all of the partners had to justify prioritizing funds for the Center within their respective organizations. This created a financial stake in the organization and outcomes of the Center that promoted active involvement in the application-writing process and the subsequent organization and leadership of the Center. In this way, the drivers of knowledge co-production were directly linked to complex urban problems under the mandates of the Consortium partners. Such problems include growing segregation and gentrification, environmental issues such as climate change adaptation in low-lying central areas, and the need to promote local and regional economic growth.

The second strategy, built directly upon the first, was joint leadership and ownership. Here the requirements for co-financing were translated into joint leadership processes where all of the public and research organizations shared in the decision-making and management of the Center from the very start of the application-writing process until the establishment of the Gothenburg Consortium after the bid was won. Joint leadership was further operationalized by recruiting Center staff from both practice and research, as well as through the co-leadership of projects, where practitioners and researchers were jointly responsible for project development; data collection, analysis, and communication; and implementation of project results. The Center staff also includes the coordinators for the Consortium partners who meet regularly with Center leadership to discuss strategic, operational, and practical issues.

The final strategy to ensure a high degree of commitment and responsibility was to create a new philosophy of working together between the partners both within the Center management and leadership, and in the Center projects and activities. This final strategy sums up the philosophy of the Center, namely that "we are all knowledge bearers, producers, and users." This philosophy is based upon the fact that the strength of the Center is its partners.

They all represent crucial knowledge and expertise from all of the governance levels in the region as well as from three research and educational institutions, including the technical, natural sciences, and social sciences, as well as the humanities. The public partners also hold the political and administrative mandates for urban change. The potential to engage, capture, and integrate this breadth of knowledge, approaches, expertise, practical know-how, and political power is what gives the Center a unique potential to contribute to creating more sustainable urban futures in the Gothenburg region. Creating commitment and responsibility thus also includes actively empowering, entitling, and encouraging different actors to engage in processes that are normally outside their realm of activities. All of the strategies outlined above were designed to tap into this potential by promoting more symmetrical responsibility, commitment, and participation in the Center activities by all of the partners, both public and academic.

Promoting co-production through projects, joint activities, and communication of results

Among the most important activities during the first two years of the Center (2010–2011) were five pilot projects that were initiated by the process leadership group during the application-writing phase. Each partner initiated or supported one or several of the pilots. For example, GR promoted a project on capturing lessons learned from the many examples of multilevel governance in the region, the City promoted a project on climate change adaptation, and VGR and IVL promoted a project on business driven sustainable urban development (Elias *et al.* 2011; Eriksson *et al.* 2011; Polk *et al.* 2013; Roth *et al.* 2011). The City, GR, and the universities together initiated two additional projects on urban empowerment and urban games (Billger *et al.* 2012; Stenberg *et al.* 2011).

These pilot projects were framed as transdisciplinary research and were managed within a co-production framework (Polk 2015). This co-production framework supported the inclusion of practice-based and disciplinary approaches from the project partners through focusing on in-depth collaboration and integration of different types of knowledge and expertise throughout the research process including problem identification and formulation, as well as project design, data collection, analysis, and preparation of results (Polk *et al.* 2013, 2014b; Westberg *et al.* 2012). Examples of methods include co-leadership of the projects by both practitioners and researchers, mixed project working groups, democratic meeting methods and a wide use of dialogue-based methods such as focus groups, workshops, and interviews. The framework also encouraged reflexive learning processes, which included discussing and reflecting upon the choices made regarding the choice and integration of different types of knowledge and expertise in each step of the projects, and a continuous focus on the usability of the results for different target groups, both scientific and policy based (ibid.). To support this work,

Center leadership created a project leadership forum for mutual learning where the project co-leaders met monthly with the leadership to exchange experiences and discuss the specific challenges they encountered with co-production in the various stages of their projects.

Managing the pilot projects went hand in hand with developing and evaluating transdisciplinary research methods through the co-production framework briefly described above. This included five rounds of interviews with the project leaders and working groups, both to gather material for analyzing the approach and to serve as forums for formative evaluations. (See Chapter 7 for more details regarding this work.) Center staff wrote a *Manual of Joint Knowledge Production* that presents a framework for knowledge co-production by outlining some of the practical problems that can arise and offering suggestions for addressing them (Westberg *et al.* 2012). A number of scientific articles evaluate the strengths and weaknesses of the co-production approach developed and tested in Gothenburg (Polk 2014, 2015).

In addition to conducting research, the Center has hosted a number of activities and events to mediate practitioner–researcher collaboration and to communicate the results of Center projects to the Consortium partners and to a wider audience. To promote collaboration, activities were organized to create local and regional meeting places for practitioners and researchers. These include targeted research seminars, speed dating, and dialogue workshops. The Consortium coordinators have also set up a number of committees for developing projects within different topical areas like segregation. Finally, the Center has actively worked to develop various kinds of tools for dialogue and visualization that have been used in participatory planning processes and educational contexts locally, regionally, and nationally (Billger *et al.* 2012). Due to the breadth and diversity of the Center's partners as well as its target groups, it has proved highly problematic to package and communicate the results of Center projects. The Consortium partners range in size from having 50 to 51,000 employees. Disseminating results to politicians, civil servants, researchers, and the general public requires tailoring the language, content, and form of the information to the different groups as well as identifying legitimate channels and forums, all of which are highly time- and work-intensive tasks.

Overall, the Center results have been communicated via several different channels. The first and initially perhaps the most important of these has been to establish the Center as a visible actor within urban planning for both practitioners, academics and the general public. Increasing visibility through branding and various communication strategies was the crucial first step in engaging stakeholders in Center activities, locally, nationally, and internationally. One of the most successful ways of communicating project results and setting up meeting places for networking locally has been the monthly Urban Lunch seminar series. For an hour and a half, practitioners and researchers come together to present and listen to project presentations and discuss playful topics over lunch. These seminars are very popular and have proved to be an effective channel for communicating with the Consortium partners. Center staff has also made presentations at international events, including Rio, WUF, ICLEI, and IFHP; at interdisciplinary and transdisciplinary conferences in Europe; and at national events organized by the Center and the Swedish National Delegation for Sustainable Cities, as well as at the Almedalen Week, a yearly national political meeting on the island of Gotland. Results of Center projects are also presented at a variety of academic conferences throughout the world. The Center has a web page, newsletters, a twitter channel, and a publication series including both academic papers and policy reports (see Center web pages).

Lessons learned from co-production in the Gothenburg region

Mistra Urban Futures is a concrete example of an attempt to create a lasting institutional arena for knowledge co-production that can contribute to bridging the gaps between research and practice. The different knowledge processes and learning activities at the Center are already influencing and being incorporated into the urban and regional development activities of public bodies in western Sweden, as well as into substantive research on sustainable urban development. A long-term institutionalization of the Center is possible because the Center's funding sources and decision-making structures include public bodies, research institutions, and external funders, all of whom are on an equal footing and have long-term commitments to the Center. Importantly, one of the demands of the external funders is that the knowledge produced be relevant and applicable for both practice and science. Establishing institutional structures and collaborative processes that meet such demands are, however, hard to achieve, despite the fact that formal structures and decisions are in place to support active participation at the Center by all Consortium partners. Some of the main barriers that exist are related to the complexity of the Center, to the number of different organizations, mandates, and areas of activity and expertise that are involved. Such diverse organizations have different ways of conceptualizing and framing their approaches to sustainable urban development as well as different institutional structures and cultures within which they work. Each organization brings with it political goals and procedures regarding decision-making, norms for developing and participating in projects, and specific mandates for change. These organizational differences are mirrored in the challenges experienced by the respective individuals when participating in the Center's joint knowledge-producing processes (Polk 2014).

One of the most pressing issues has been lack of time and resources. Sharing responsibility for organizing the Center and identifying and formulating problems takes much more time than processes that are led by one main actor, or based on traditional research or consultation relationships. Put simply, doing things together especially across diverse scientific and policy cultures, takes more time, to get acquainted, to establish trust, reconcile differences, and navigate communicative and organizational barriers. The dual demands of producing both scientific and policy input are also difficult to integrate in practice. This is due to both the different standards and needs in science and practice, to the different forms of communication that are required, as well as to the diversity of target groups involved. Such differences mean that many of the projects and activities at the Center are messy and unwieldy. As noted above, organizational and institutional barriers and conditions limited the ability of the participants to integrate project results back into their respective organizations. Reconciling such differences proved to be a greater challenge than had been anticipated at the conception of the Center.

Finally, and as will be taken up in Chapter 7, identifying results and gauging impacts of the pilot projects and other Center activities was highly problematic and entails creating new evaluative techniques to identify the outputs as well as gauge possible outcomes. As will be also be discussed in Chapter 3, while the focus of co-production is new knowledge production, new knowledge is only one output of co-production processes. Increased understanding, insight and mutual learning regarding the complexity of urban sustainability by the involved practitioners and researchers is an equally important outcome, though much more intangible therefore hard to identify and account for. The messiness of the Center project and activities, while problematic for the reasons outlined here, also mirror the strengths of the Center. Their messiness is a result of bringing together diverse actors, spheres of activity, mandates and disciplinary cultures. It is also here that the most learning can be achieved.

Concluding remarks

The collaborative spirit of the greater Gothenburg region has laid a good foundation for efforts to develop and institutionalize various forms of researcher- and state-initiated co-production among politicians and administrators, and between research and practice. Various actors have fostered this spirit over time in their attempts to better understand and manage the critical issues, barriers, and complex problems that oppose urban growth and development. This collaborative spirit has been closely linked to national policy and regulations, not least during the forty-four-year period when a succession of Social Democratic governments built the modern welfare state and made Sweden a home for all its people (Holgersson *et al.* 2010). However, collaborative processes that co-produce relevant knowledge are only one small piece of the sustainability puzzle. There is also a need for new types of coordinated action by diverse actors, stakeholders, and the public to implement such knowledge in complex processes of social change. In this, a number of barriers and difficulties remain to be addressed and managed (Delegationen för Hållbara Städer 2012). These include administrative and organizational boundaries, established practice and sector responsibilities, as well as competing understandings of the challenges society is currently facing and how they should be addressed. Studies of political and administrative processes for visioning, consultation, and urban planning show a continued dominance of a neoliberal sustainability agenda where stimulating further economic growth is the dominant policy and planning approach to addressing environmental and social problems (Cullberg *et al.* 2014; Polk 2010). This neoliberal agenda is sometimes challenged but never defeated by a more transformative agenda, where the economy is seen instead as the means to reach the goal of sustainability, namely global justice within the frame of environmental carrying capacity. Despite the examples of different emphasis on social and environmental problems, much of the policy, planning, and actions taken in the Gothenburg region are still embedded in this status quo agenda. So far, however, the focus on economic growth in the region has neither been able to prevent environmental degradation and increasing resource use nor slow down growing social polarization and segregation. It is in this context of the failure of neoliberal approaches to solve pressing urban problems, that new ways of working together are attracting growing interest from the political and administrative organizations in the region. This is the institutional context where co-production has gained a foothold in planning, policy making, and research in Gothenburg.

References

Andersson, G. I., Fredman, P., Holmberg, J., Hulthén, A., Markides, K., Ransgård, J., Trové, J. and Törsäter, B. (2011) 'Låt Sverige bli en framtida test arena' [Let Sweden be a future test arena], *Göteborgs Posten*, January 18.

Andersson, P. (1993) *Sveriges kommunindelning 1863–1993*. Mjölby: Draking.

Billger, M., Alfredsson, K., Lindkvist, J., Myren, P. and Clase, K. (2012) *Urbana spel* [Urban games]. Mistra Urban Futures Working Paper. Göteborg, Sweden: Mistra Urban Futures.

Chalmers (2010) *Environmental Science and Sustainable Development*. Downloaded March 15, 2014 from www.chalmers.se/en/research/environmental-science-and-sustainable-development/Pages/default.aspx.

Cullberg, M., Montin, S. and Tahvlizadeh, N. (2014) *Urban Challenges, Policy, and Action in Gothenburg: GAPS Project Baseline Study*. Mistra Urban Futures Reports, 5.

Delegationen för Hållbara Städer (2012) *Femton hinder för hållbar stadsutveckling* [Fifteen barriers to sustainable urban development]. SOU M 2011:01/2012/66.

Elias, K., Forsemalm, J., Gustafsson, J., Johansson, E., Johansson, M., Lilled, L., Montin, S. and Löf, Y. (2011) *Governance: att styra med insikt snarare än avsikt. En studie om flernivåstyrning och sektorssamverkan i Göteborg och Malmö* [Governance: to govern by insight instead of by intent, a study of multilevel governance and cross-sector collaboration in Göteborg and Malmö]. Mistra Urban Futures Policy Report. Göteborg, Sweden: Mistra Urban Futures.

Eriksson, E., Sandoff, A., Värmby, G., Rönnborg, P., Jensen, C., Molnar, S., Bergstrand, B.O., Antonsson, A.B-, Elias, K., Kildsgaard, I., Bolin, L. and Wolf, C. (2011) *Affärsdriven hållbar stadsutveckling: Goda exempel och potential* [Business-driven sustainable urban development: Good examples and potential]. Mistra Urban Futures Policy Project Report. Göteborg, Sweden: Mistra Urban Futures.

Etzkowitz, H. and Lweydesdorff, L. (2000) 'The Dynamics of Innovation: From National Systems and "Mode 2" to a Triple Helix of University–Industry–Government Relations', *Research Policy*, 29: 109–23.

Falkenmark, G. (2010) 'The Gothenburg Spirit – From Donations to Networking', in H. Holgersson, C. Thörn, H. Thörn and M. Wahlström (eds), *(Re)searching Gothenburg – Essays on a Changing City*. Hässleholm: Glänta Production.

Five Clusters (2012) 'Five Clusters in west Sweden with strength and potential for the future', information brochure. Gothenburg: Five Clusters.

GPL (1959) *Generalplan för Göteborg 1959* [Masterplan for Gothenburg 1959]. Göteborg: Staden.

GR (2006) *Uthållig tillväxt. Mål och strategier med fokus på hållbar regional struktur* [Sustainable growth – Goals and strategies with a focus on sustainable regional structures]. Göteborg: Göteborgsregionens kommunalförbund (GR). Available at: www.grkom.se.

——(2008) *Strukturbild för Göteborgsregionen. En överenskommelse att vi gemensamt tar ansvar för att den regionala strukturen är långsiktigt hållbart* [Structural illustration for the Gothenburg region]. Göteborg: Göteborgsregionens kommunalförbund (GR). Available at: www.grkom.se.

Hansson, J. (1999) *Millennium – årtusendets bok*, Göteborgsposten.

Holgersson, H., Thörn, C., Thörn H., and Wahlström, M. (eds) (2010) *(Re)searching Gothenburg – Essays on a Changing City*. Hässleholm: Glänta Production.

Holmberg, J. (2012) 'Sustainable Development as a Driver for University Transformation', in Global University Network for Innovation (GUNI) series on societal commitment of universities, *Higher Education in the World 4. Higher Education's Commitment to Sustainability: From Understanding to Action*. Basingstoke: Palgrave Macmillan.

Jörnmark, J. (2005) *Göteborgs globalisering*. Göteborg: Göteborgs Stadsmuseum.

Khakee, A. (2010) 'Assessing Institutional Capital Building in a Local Agenda 21 Process in Göteborg', *Planning, Theory & Practice*, 3(1): 53–68.

Lundqvist, L. (2000) 'Capacity-building or Social Construction? Explaining Sweden's Shift towards Ecological Modernisation', *Geoforum*, 31: 21–32.

Montin, S., Tahvilzadeh, N., Cullberg, M. and Stark, E. (2014) *En analys av hållbarhetspolitiken i Göteborgsregionen* [An analysis of the sustainability policies in the Gothenburg region], draft March 2014.

ÖP (1993) *Översiktsplan för Göteborg 1993* [Comprehensive Plan of Gothenburg 1993]. Göteborg: Göteborgsstad.

——(1999) *Översiktsplan för Göteborg 1999* [Comprehensive Plan of Gothenburg 1999]. Göteborg: Göteborgsstad.

Peterson, M. (2010) 'Four Centuries of Multiculture', in H. Holgersson, C. Thörn, H. Thörn and M. Wahlström (eds), *(Re)searching Gothenburg – Essays on a Changing City*. Hässleholm: Glänta Production.

Polk, M. (2010) 'Sustainability in Practice: The Interpretation of Sustainable Development in a Regional Planning Arena for Dialogue and Learning in Western Sweden', *Planning Theory & Practice*, 11(4): 473–89.

——(2011) 'Institutional Capacity Building in Urban Planning and Policy Making for Sustainable Development: Success or Failure?', *Planning Practice & Research*, 26(2): 185–206.

——(2014) 'Achieving the Promise of Transdisciplinarity: A Critical Exploration of the Relationship between Transdisciplinary Research and Societal Problems Solving', *Sustainability Science*, 9: 439–451.

——(2015) 'Transdisciplinary co-production: designing and testing a transdisciplinary research framework for societal problem solving', *Futures* (in press).

Polk, M., Kain, J. H. and Holmberg, J. (2013) 'Mistra Urban Futures: A Living Laboratory for Urban Transformations', in A. König (ed.), *Regenerative Sustainable Development of Universities and Cities: The Role of Living Laboratories*. Cheltenham: Edward Elgar.

Polk, M. and Malbert, B. (2010) *Samverkan och förankring inom hållbar stadsplanering: Erfarenheter från K2020, Kollektivtrafikprogrammet för Göteborgsregionen* [Cooperation and commitment building within urban sustainability planning: Experience from K2020, the public transport program for the Gothenburg region]. Report to the Swedish Transport Administration (Vägverket).

Polk, M., Malbert, B. and Kain, J.-H. (2009) *The Göteborg Center of Excellence for Sustainable Urban Futures*. Göteborg: Mistra Center Application.

Poulsen, R. and Sornn-Friese, H. (2011) 'Downfall Delayed: Danish Shipbuilding and Industrial Dislocation', *Business History*, 53(4): 557–82.

Roth, S., Thörn, P., Buhr, K., Moback, U., Morrison, G., Knutsson, P. and Areslätt, H. (2011) *Frihamnen i ett förändrat klimat Klimatanpassningsstrategiers påverkan på hållbar utveckling* [The influence of climate adaptation strategies on sustainable development]. Mistra Urban Futures Pilot Project Report. Göteborg, Sweden: Mistra Urban Futures.

SCB (2014) Downloaded April 20, 2014 from: www.ssd.scb.se.

Soidre, T. (2010) 'Good Housing – for All?', in H. Holgersson, C. Thörn, H. Thörn and M. Wahlström (eds), *(Re)searching Gothenburg – Essays on a Changing City*. Hässleholm: Glänta Production.

Stenberg, J., Fryk, L., Bolin, E., Borg, P., Castell, P., Evenås, U. and Larberg, V. (2011) *Urban Empowerment: Cultures of Participation and Learning*. Mistra Urban Futures Working Paper, Göteborg, Sweden: Mistra Urban Futures.

Swedish Institute (2013) *Sweden's Environmental Policy*. Downloaded May 3, 2014 from: sweden.se/nature/environmental-policy.

Thörn, C. (2010) 'Fishing for Authenticity – the City as Theatre', in H. Holgersson, C. Thörn, H. Thörn and M. Wahlström (eds), *(Re)searching Gothenburg – Essays on a Changing City*. Hässleholm: Glänta Production.

Thörn, H. (2010) 'Proletarians, Proggers and Punks – Folkhemmet, Haga and a Hundred Years of Social Movements', in H. Holgersson, C. Thörn, H. Thörn and M. Wahlström (eds), *(Re)searching Gothenburg – Essays on a Changing City*. Hässleholm: Glänta Production.

VGR (2005) *Visionen: Det goda livet* [Vision: The good life]. Göteborg: Västra Götalandsregionen.

Warburg, K. (1913) *Viktor Rydberg, hans levnad och diktning*. Stockholm: Albert Bonniers Förlag.

Wärmby, G. (1992) 'Dagens spelplan – Ett samhällsperspektiv', in B. Malbert (ed.), *Ekologiska utgångspunkter för planering och byggande*, Efem arkitektkontor, Byggforskningsrådet T35:1992.

Westberg, L., Polk, M. and Frid, A. (2012) *Manual of Joint Knowledge Production for Urban Change*. Mistra Urban Futures Report. Göteborg: Mistra Urban Futures.

Web pages

Chalmers web pages www.chalmers.se.
FORMAS web pages www.formas.se.
K2020 web pages www.k2020.se.
Mistra web pages www.mistra.se.
Mistra Urban Futures web pages www.mistraurbanfutures.org.
S2020 web pages www.socialhallbarhet.se.
Swedish Scientific Research Council (Vetenskapsrådet) web pages www.vr.se.
VINNOVA web pages www.vinnova.se.

3 Co-producing knowledge for whom, and to what end?

Reflections from the African Centre for Cities in Cape Town

Warren Smit, Mary Lawhon, and Zarina Patel

Introduction

Cape Town, South Africa, is a city faced by a range of sustainability challenges. It is home to unique biodiversity and is a well-known tourist destination. However, it is also a very inequitable and segregated city, with extremes of poverty and wealth, and faces a range of biophysical, social, and economic challenges. There is a growing realization in South Africa, and Cape Town in particular, that no one field of "knowing" is sufficient to both understand the multidimensionality of urban problems and deliver solutions and approaches for complex situations. While knowledge is not the only barrier to urban sustainability, it is an important one.

In order to help fill this gap in understanding, the African Centre for Cities (ACC) at the University of Cape Town seeks to develop progressive research methodologies through which to connect academia, policy, and practice. It draws its inspiration from the growing push for research that transcends or works between disciplinary boundaries and connects with real-world problems, as well as from the need for theoretical developments based on empirical findings in the global South. However, research that attempts to impact on real-world problems needs to be normative, and it is necessary for researchers to clearly articulate the norms that inform the research process and outcomes.

In this chapter, we reflect on the experiences of co-production in the ACC, with a specific focus on the methods used and the challenges that underpin such normative research. We seek to draw out lessons from previous experiences, and show how they matter in new research. The chapter discusses the CityLab program, ACC's flagship co-production program since 2008, which consists of a number of interdisciplinary platforms dedicated to thematic research areas based in and focusing on Cape Town. There is also some reflection on the more recent Knowledge Transfer Program, a collaboration between ACC and the City of Cape Town which began in 2012 as part of the Mistra Urban Futures program. This chapter is informed by the authors' experience of co-production within these two programs, commentary from other participants, as well as a one-day workshop conducted in July 2012 and attended by academics and City officials from both programs.

First of all, we briefly look at the concept of "co-production of knowledge" and the importance of normativity in this. Second, we examine the drivers for, and context of, knowledge co-production in Cape Town. The African Centre for Cities and its two programs for co-producing knowledge on Cape Town (the CityLab program and Knowledge Transfer Program) are then described. Finally, four issues relating to the importance of normativity in the co-production of knowledge are reflected on, using examples from ACC's experience: defining the research question/problem; identifying the intended impacts; choosing research methods; and deciding who is involved and how. These case studies of the ACC's CityLab Program and Knowledge Transfer Program show that, while there is no single approach to the co-production of knowledge at the ACC, it is extremely valuable to work with multiple partners to generate more appropriate knowledge to shape urban sustainability outcomes. The examples discussed in this chapter highlight that in order to have successful knowledge co-production it is essential that all partners are explicit about their normative assumptions and goals.

Normativity and the co-production of knowledge

Addressing the importance of connecting research with real-world outcomes has become an increasing focus of universities over the last decade (as noted in Chapter 1). The motivations for this new kind of knowledge are diverse, and different literatures label and describe these motivations in slightly different ways; see, for example, Hessels and van Lente (2008) on Mode 2 knowledge production, Jasanoff (2004) on co-production, Klein (2004) on transdisciplinarity, and Petts *et al.* (2008) on interdisciplinarity. However, two main threads are common in these literatures. First, there is a sense that traditional disciplines do not recognize integration and complexity, and therefore do not produce knowledge which is of much practical use. One view is that "[t]he main trap of disciplinary thinking is the *uncritical* acceptance of specific methods and analytic tools, irrespectively of the specific features of the objects under investigation" (Leonelli 2005: 2).

Second, there is a realization that knowledge is subjective, and that researchers need to engage more directly with understanding its situated construction. In the twentieth century, there was a paradigm shift in philosophy toward looking at reality as a socially constructed phenomenon. The concept of the co-production of knowledge is based on this understanding that knowledge is subjective and socially constructed. For example, Jasanoff argues that:

> the ways in which we know and represent the world (both nature and society) are inseparable from the ways in which we choose to live in it … [S]ociety cannot function without knowledge any more than knowledge can exist without appropriate social supports. Scientific knowledge, in particular, is not a transcendent mirror of reality. It both embeds and is

> embedded in social practices, identities, norms, conventions, discourses, instruments and institutions.
>
> (2004: 3)

However, there is rather limited articulation of the relationship between knowledge and power in this literature. Science studies have suggested that we must understand not only that knowledge is subjective and situated, but also that it constructs, is constructed by, and is used to shape power relations. As Jasanoff (2004) notes, much can be learned about the co-production of knowledge through a more careful integration of science studies; see also St Clair (2006), Mitlin (2008), Pestre (2003), and Pohl *et al.* (2010).

There is some uncertainty about the appropriate terminology to describe the different kinds of work that engage in research attempting to cut across disciplines and sectors and engage with real-world problems. Here, we use "interdisciplinary" to refer to work that cuts across the boundaries of traditional academic disciplines, and we use "co-production of knowledge" for processes that bring together academic researchers and practitioners (both in government and elsewhere) to produce knowledge that is useful for policy, practice, and academia. While acknowledging the importance of clarity, we seek in this chapter to move beyond debates on motivation and terminology. Like many others, we suggest the importance of better understanding the challenges and opportunities of such work in practice. There is a small but growing number of case studies of attempts to cross disciplines and co-produce knowledge (for example, Anderson *et al.* 2013; Carolan 2007; Goebel *et al.* 2010; Pohl 2008).

Making more specific generalizations remains a challenge, in part because there are not enough published discussions of interdisciplinarity and co-production in practice. However, more significantly, it is the context-based subjectivity of doing this type of work that makes generalization difficult (Lemos and Morehouse 2005). This differs among disciplines (interdisciplinary natural science is reportedly rarer than interdisciplinary social science) as well as contexts; for example, national differences have also been noted (Shinn 2002; Tuanneinen 2005). While many have encouraged reflexivity to improve research methods, we suggest that there is also a need for a specific type of reflexivity which is necessary for understanding the particular processes and outcomes of co-production: consideration of the normativity of knowledge and change. Normative statements and theories are about how things ought (or ought not) to be, as opposed to descriptive statements and theories about what is (or is not). Normativity, which is the (subjective) view that the world should be different in some way to how it currently is, is usually contrasted with positivism, the traditional academic approach (especially in the natural sciences), which attempts to describe reality and find the "truth" in an "objective" and "unbiased" way.

Some academic disciplines are inherently normative, for instance urban planning and public health, as they actively try to change the world in specific ways. Most academic disciplines are not explicitly normative, but are usually implicitly so, as their practices are based on assumptions about the way the world should be. For example, Pickersgill (2012: 599) notes of neuroscience research that "knowledge is partially constituted through particular regimes of normativity consisting of formal ethical review, informal moral discourse, official/unofficial protocols within specific scientific communities, and individual personal beliefs and emotions." Further, the selection of methodology is a normative process. As Leonelli suggests:

> [T]he representation and analysis of social realities require a normative choice among several available tools, which has to be justified according to the interests and objectives of the researchers (whether epistemological, ethical or else) … [T]he recognition of pluralism in research tools and methods, which is arguably a core virtue of interdisciplinary approaches, calls for a heightened awareness of the normative consequences of choosing specific tools and perspectives for inquiry.
>
> (2005: 3–4)

It is particularly important to be explicit about normativity in knowledge co-production, where it is intended to grapple with, and impact on real-world problems, and where people from different disciplines and sectors might have very different norms, values, and ethics. First, change is not a neutral concept. The desire and ability to successfully create change have, we suggest, been valorized in the literature, with insufficient attention being paid to what should be changed, in what way, and for whom. Responding to this point, Flyvbjerg (2004: 289–90) suggests four key questions for urban planning that apply more generally: "(1) Where are we going with planning? (2) Who gains and who loses, and by which mechanisms of power? (3) Is this development desirable? (4) What, if anything, should we do about it?" The South African case again provides a useful context in which to reiterate the ambiguity of the academic–policy–practice nexus because there is a long history of intervention and engagement, providing both critique of apartheid and intellectual justification for it (Maharaj and Narsia 2002; Sanders 2002). Thus, intervention is not inherently positive for everyone, and the ideology underpinning the type of real-world impact needs to be clearly identified in any such work.

Second, we suggest that stepping outside the disciplinary boundaries blurs the theoretical identifiers often present in disciplinary academic literature. Scholarly training helps academics work through the ideological underpinnings of certain key theories and associated authors. Who one cites helps identify one's ideological position to an intellectual community generally familiar with the cues this suggests. Of course this is not to say that one agrees with everyone cited, but the citations are generally used to signify intellectual lineage. For example, in geography, there are significant if somewhat subtle differences between citing land change science, vulnerability frameworks, and political ecologists. Phrases such as "historical materialism" and "the right to the city" indicate particular lineages of urban studies to urbanists, but may not be intelligible to lay readers or others from other fields. Moving outside of traditional academic boundaries and into co-production, these signifiers are likely to be less clear, as both those involved in the research process and the audience may be unfamiliar with many of the key concepts, authors, and signifiers.

Therefore, methodological rigor in a co-production process requires more than just reflexivity. To understand how knowledge is co-produced, including both the process and outcomes, requires not only acknowledging the subjectivity of decisions, but explicitly reflecting on what informs the co-production process. For example, who is participating? What informs a decision when a challenge arises? What changes and impacts are desired, by whom, and why? Factors informing the process include the different positionalities of participants (who they are as a result of their birth, upbringing, education, occupation, and other life experiences). Related to this are the different ideologies, values, views on what we can know (ontology), and views of how we can know (epistemology) of participants. Ontologies and epistemologies are usually strongly influenced by the disciplinary background of participants, and place various constraints on how knowledge is generated and analyzed, as well as the way in which knowledge is used in nonacademic settings. To be able to begin developing a clearer understanding of how co-production actually happens, we point to the need to be explicit about positionalities, ideologies, values, ontologies, epistemologies, theoretical perspectives, and methodological approaches of participants. Only by putting this on the table can we begin comparing and learning from different examples of the co-production of knowledge across sites and subjects.

Drivers for co-production

The local drivers for knowledge co-production vary across contexts. In South Africa, the transition to a democratic society over the past few decades, combined with enormous challenges inherited from the past and a growing skills shortage, has created conditions that have particularly stimulated attempts at co-production.

In 1994, the new democratic government of South Africa inherited a host of challenges. South African cities were highly inequitable and segregated, had large and growing numbers of unemployed people, and significant numbers of people living in informal settlements without adequate access to services. Issues of sustainability also increasingly began to be put on the agenda during the 1990s (Lawhon and Patel 2013).

While scholars have a limited understanding of the nature of the complex problems facing South African cities, the state too is struggling to keep pace with policy responses. The political changes which have occurred in the past twenty years have provided opportunities for change, including the development of new legislation in almost every key sector, often drawing on international good practice.

Policy changes post-1994 have included the concept of new forms of knowledge production, most notably in the 1997 White Paper on Higher Education (Jansen 2002). Boersma *et al.* argue that, in addition to transformation of staff and mergers occurring in tertiary institutions, "most scholars have interpreted change in South Africa's higher education system against the background of what has been called the new mode of knowledge production … This idea became known as Mode 2; it presupposes inter- and trans-disciplinary research and education, growing importance of university–industry relationships and context-driven, problem-focused research and education" (2008: 212).

However, the notion of extra-academy engagement in South Africa—including the developmentalist agenda which has linked research and practice—certainly preceded this moment. South Africa has long followed a model of "knowledge inputs" into policy processes from consultants and academics in a context where a capacity deficit is the norm. Indeed, in contrast to the criticism that research needs to be made relevant, South African urban researchers have been criticized for overemphasizing practical work and doing too little theoretical reflection (Hammett 2012; Parnell 2007; Todes 2006). It is certainly not our intention to overstate the prevalence of engaged research in South Africa, but we do wish to point to the long history of research–practice connections as a wider lens through which to view current trends (Ernstson *et al.* 2014).

South African cities, we suggest, provide a useful example of the blurring of the boundaries between theory, practice, and civic engagement. As Parnell says of geography, but which also applies more widely in the social sciences, "consulting, policy or applied research is a ubiquitous feature" (2007: 111). In part this relates to the moral desire for engagement during the apartheid era. As Oldfield *et al.* (2004) recount, apartheid-era civil society groups sought intellectual support from universities during the 1980s, resulting in long-term cooperation and the co-production of knowledge. The upshot of these academy–civic partnerships, born through "sustained and critical engagement," was the creation of "conditions for intellectual creativities in the recipient NGO environment, in community-based organizations (CBOs) and in university contexts" (Oldfield *et al.* 2004: 292). However, as noted above, academic connections also existed which promoted the legitimation of apartheid, highlighting the need to explicitly consider the normativity of such connections.

The context of Cape Town

Cape Town is a particularly challenging, but potentially revelatory context in which to examine the co-production of knowledge. South Africa, and Cape Town in particular, may be considered middle income or transitional, but this belies the extensive inequality of the city. Cape Town, which had a population of 3.7 million people in 2011 (City of Cape Town 2012), is one of the most inequitable cities in the world (UN-HABITAT 2010). About 40 percent of households are classified as poor, in other words as having insufficient income to access basic necessities such as food and shelter (City of Cape Town 2011). In addition, as with other South African cities, Cape Town is characterized by an inefficient, fragmented spatial structure that preserves racial inequalities inherited from the apartheid period, and post-apartheid spatial changes are exacerbating this fragmentation and segregation (Lemanski 2007; McDonald 2008; Turok 2001; Wilkinson 2000). This inequality presents particular challenges for governance and sustainability, as different constituents have different desires and expectations from government, academia, and other stakeholders who intervene in the city. Global environmental change and resource depletion are exacerbating these challenges. While South Africa is generally considered to have a progressive national environmental policy, the Cape Town city–region has the most developed local environmental policy (Crane and Swilling 2008). Despite this policy context, however, Cape Town remains an unsustainable city (Swilling 2010).

In Cape Town, there has been a tension between, on the one hand, a push in the city to become a globally competitive world city, to position itself as a global leader in tourism, design, and technology (Lemanski 2007; McDonald 2008; Wilkinson 2004), and, on the other hand, the competing pressure for service delivery, job creation, poverty alleviation, and overcoming the historical legacy of inequality. These two opposing pressures often work against each other, drawing resources in different and often contradictory ways. As Wilkinson (2004: 228) notes, "It seems self-evident that commitments such as that to alleviating poverty will be likely to sit somewhat uneasily alongside, for instance, the commitment to promoting tourism as development priorities and that, inevitably, some trading-off between them will be required." While there is certainly awareness of this tension among City officials, there is no clear guidance from the local government on how to respond to it.

Additionally, there are political differences which make Cape Town unusual in South Africa. In contrast to the rest of the country, where since the transition to democracy in the mid 1990s provincial government and local government have almost everywhere been led by the African National Congress (ANC), the dominant liberation party, the City of Cape Town and the Western Cape Province have both seen relatively frequent changes in power, and the oppositional Democratic Alliance (DA) is currently in power in both the City and the Province.[1] Many commentators have noted that the policies followed by the DA and ANC have not been radically different,[2] but the high levels of political contestation and relatively frequent changes of administration have resulted in a unique institutional environment with a "lack of trust between the political leadership and top officials, especially where they were appointed by the other party; a highly partisan political environment, where the dominant party typically sought little positive engagement with the opposition; and a strong sense of insecurity, which manifested itself in a strong assertion by political leadership of their power prerogatives" (Schmidt 2008: 123–4).

This political scenario creates challenges and opportunities, and a political environment where experimentation has both higher risks and rewards. Cape Town also serves as a useful site because of the intellectual work occurring in and focusing on the region. The wider metropolitan region is home to four universities,[3] and this quantity, quality, and concentration of academics makes it an intellectual hub on the continent. The connections between the University of Cape Town (UCT) and the municipal government are relatively close; these relationships have been facilitated by the presence of graduates from the university in various City departments.

In such a diverse context it is essential to examine and critically reflect upon knowledge and its production, use, and politics. No single group can solve these complex urban problems, as a range of knowledges and expertise is needed to inform policy and decision-making, as well as in theorizing and describing the city. There is no single, unifying vision for the city, and the municipal government's many goals compete for attention and resources. There is therefore a need to reflect on the normative implications of work which seeks to bring together diverse stakeholders to change policy and practice. Further, there is a need to consider the role of knowledge in facilitating or creating change—how it is used, by whom, and to what end. The City has recognized the need to improve knowledge for urban governance, as is evident in their partnerships with academic institutions such as the African Center for Cities.

Developing co-production practice at the African Center for Cities

The ACC was established in 2007 at the University of Cape Town (UCT) as an interdisciplinary research institute on urban issues. The ACC emerged from "a committed network of scholars at UCT who were keen to see an interdisciplinary focus on urban studies" be revived at the university (Pieterse 2013: 27). The catalyst for establishing the ACC was the development of an interdisciplinary master's program aimed at mid-career built-environment professionals, and it was decided that this program "should be embedded in a lively interdisciplinary research centre" (Pieterse 2013: 27). Funding was then raised from a wide range of local and national sources, the Director was appointed in 2007, and the appointment of additional employees followed. ACC's mission is "to facilitate critical urban research and policy discourses for the promotion of vibrant, democratic and sustainable urban development in the global South from an African perspective" (ACC website). The research agenda of ACC is "structured around the problematic of sustainable human settlements, i.e., what they are and how to bring these universal rights into being in complex political, economic and institutional settings" (Parnell *et al.* 2009: 236).

As a platform for new ways of doing urban research, the ACC "gathers together scientists, engineers, planners and social scientists to work collaboratively on the problem of rethinking the city … to produce new knowledge …

to aid interventions by the state, citizens and their organizations and the private sector" (Parnell *et al.* 2009: 241). The focus of the ACC's work is to engage with policy and practice and develop improved methods for doing such work. The ACC seeks to provide an African lens on the urban context, developing both empirical work and theorizations based on this context. The ACC's explicit intention to both understand and change the city clearly makes it a normative entity, differing somewhat from more traditional academic and research institutions. In particular, the promotion of social justice is a core theme.

The CityLab program and the Knowledge Transfer Program (KTP) both play an important role in ACC's facilitation of critical urban research and policy discourses in Cape Town. These two programs are discussed below.

The CityLab Program

ACC's first research program on Cape Town, the CityLab program, was initiated during 2008 as an interdisciplinary applied research program for learning about sustainable urban development from the experiences of Cape Town. Through engagement between the ACC and various partners in UCT, the City of Cape Town, and the Western Cape Provincial Government, a number of key CityLab topics were identified, and researchers were then recruited to work on these themes. In all, five CityLabs were started during 2008 and 2009: the Central City CityLab; the Philippi CityLab; the Climate Change CityLab; the Urban Flooding CityLab; the Urban Ecology CityLab; and the Healthy Cities CityLab (see Table 3.1 for more details on the first round of CityLabs). The CityLabs are all intended to have finite lifespans, and as some CityLabs have come to an end, new ones have been started.

The CityLabs attempt to move beyond disciplinary knowledge and draw insights from across the university (Anderson *et al.* 2013). Additionally, they have worked and continue to work with stakeholders including government, communities, NGOs, business, and the university. Over the course of their operation, multiple engagements helped to develop relationships, "social learnings," and shared expertise, often prompted by concerns within government (Anderson *et al.* 2013). Increasingly, ACC researchers have adopted the term "co-production of knowledge" to describe the phenomenon of generating new knowledge through these processes.

As noted on the ACC's website, the work of the CityLab program "is characterized by: a focus on empirical research on the Cape Town city region; bringing together researchers and practitioners from different disciplines and professions; learning from practice and dissemination of knowledge that will build sustainable human settlements." Here we see the interdisciplinary focus and commitment to co-production, as well as a particular normative ambition: sustainable human settlements. Although the term "CityLab" suggests an intention of "generating sufficient knowledge through repeated endeavour, observation, measurement, monitoring, and engagement to inform a robust

Table 3.1 The African Centre for Cities' first round of CityLabs

CityLab name and time frame	*Focus*	*Format*	*Publication output*
Central City CityLab, 2008–14	Toward an expanded understanding of the complex debates related to the densification of the Central City to a wider audience. This was achieved by conducting new research, and through using the collected material from seminars.	*Seminar series* Open to public, held at University, attended by a mix of academics, practitioners, and government officials.	Book publication (to be published in 2014)
Climate Change CityLab, 2009–12	Toward a common platform for people of different disciplines and backgrounds, working on climate change at the city scale. The intention was that the resultant knowledge community would enable an enhanced understanding of local climate risks and adaptation options.	*Think tank* By invitation, hosted alternately by the City of Cape Town and ACC. Plenary meetings as well as smaller working group meetings of various disciplines and sectors of society.	Book publication (published in 2012)
Healthy Cities CityLab, 2009–present	Toward a long-term, phased, research project with initial interdisciplinary research on the relationship between the physical urban environment and human health.	*Research project* Project meetings open only to predefined working group comprised of academics from different disciplines. Meetings held at the University. Later phases involve engagement with policy-makers and practitioners to create healthier urban environments.	Book chapter (published in 2014) and academic journal articles (in press)
Philippi CityLab, 2008–14	Toward exploring and debating the preconditions for sustainable and equitable local development in low-income areas in the city, using the Philippi area (which has a concentration of informal settlements) as a case study.	*Seminar series* Open to public, held at University as well as in Philippi, attended by a mix of academics, practitioners, and government officials.	Book publication (to be published in 2014)

Table 3.1 (continued)

CityLab name and time frame	*Focus*	*Format*	*Publication output*
Urban Ecology CityLab, 2010–14	Toward a collective understanding of the status quo of urban ecology in the City of Cape Town. To this end, unpublished work and ideas were shared, debated and then written up for publication.	*Seminar series* Open to the public, held at University, attended by a mix of academics, practitioners, and government officials with a high turnover in attendance between meetings. Meetings held at the University and at field sites around Cape Town.	Special issue academic journal (published in 2012)
Urban Flooding CityLab, 2009–13	Toward exposing researchers and practitioners to a wide range of perspectives on the causes, impacts and solutions to flooding in Cape Town and documenting different perspectives to flood management and working towards more holistic and proactive strategies of reducing flood risk.	*Research project and seminar series* Seminars and field trips open to the public, project meetings open only to predefined working group comprised primarily of academics from different disciplines. Meetings held at the University. The research work involved government officials, community organizations, and NGOs.	Report publication (published in 2013) and academic journal articles (published in 2014)

Source: Adapted from Anderson *et al.* (2013).

theory" (Anderson *et al.* 2013: 2), the actual methods employed in the CityLabs were much more open-ended and less positivist. These methods usually involved seminar series with presentations from academic researchers and practitioners from a range of disciplines and sectors, collaborative research projects, and the production of a book or special issue bringing together contributions from academics and practitioners on a specific topic. These processes were often very messy, with challenges in understanding each other's terminology and worldviews, and with the usual "micropolitics and personality clashes" that can be expected when a diverse range of people from diverse backgrounds attempt to work together (Anderson *et al.* 2013: 7). Nonetheless, the CityLabs all played an important role in bringing together different types of knowledge, building a more integrated body of knowledge on Cape Town, and enriching the work of both academic researchers and practitioners involved in the CityLabs. The CityLab program has "certainly made a contribution in the co-production and dissemination of knowledge and in some instances has clearly facilitated shifts in practice and thinking" (Anderson *et al.* 2013: 2).

Linked to the CityLabs are the "policy roundtables." These are processes through which ACC advises the City and Province on key policy issues. (These processes often take the form of informal discussions between academics and officials around a table, hence the name.) Notable examples include assisting in the development of an urbanization framework for the City and helping the Province with the development of their One Cape long-term strategy. In these processes, academic researchers from the ACC worked together with officials from the City and Province to develop policies that reflect both academic knowledge and practical knowledge of the challenges and opportunities.

The Knowledge Transfer Program

ACC launched the Knowledge Transfer Program (KTP) in July 2012, which allows for a unique way of working together for the City and University. The objective of the partnership is to develop knowledge partnerships that are better positioned to provide complex policy responses to complex urban sustainability challenges. The approach to co-production, formalized through a memorandum of understanding between the City of Cape Town and the ACC, is to develop knowledge platforms that straddle the researcher–practitioner divide. The aim of the co-production process in the KTP is for the process to benefit both officials and academics. City officials will be better positioned to provide defensible policy responses to complex urban problems while academics work with City officials both to generate data to input into policy processes and to contribute to academic debates on urban sustainability. This will help enhance our empirical and theoretical understanding of the experience of Southern cities, and Cape Town specifically.

The focus of the KTP has been on making the policy and decision-making process supporting a sustainability agenda more defensible and legible. The

KTP recognizes that enabling alternate development pathways demands the engagement of experts from ecological/technological, social, and economic disciplines. While expertise is considered essential, the need to access and assemble tactical knowledge in order to understand and address the limits and opportunities of policy responses is an added dimension in this approach (Bulkeley 2006). The focus on local government policy and decision-making allows for a more focused set of relationships in the KTP, and is achieved through a bi-directional approach. In one direction, four Ph.D. researchers from the ACC are partnered with city counterparts working on policy and implementation challenges in response to contemporary urban problems: climate change, the green economy, the space economy and densification, and energy governance. These researchers are embedded in City structures for seven months at a time over a three-year period. These compound (and complex) working relationships pair the evidence-based knowledge in which the city practitioners have expertise with the theoretical and methodological rigor of academia to provide defensible policy positions and implementation frameworks. The researchers, in turn, are actively reflecting on, and attempting to document the often opaque informal processes of policy development and decision-making in the City, thereby making policy and decision-making more accessible.

In the other direction, the City Officials Exchange Program provides selected city officials deeply involved in urban sustainability policy development an opportunity to spend two months at UCT writing up an aspect of their work with an academic writing partner. The aim is to produce a publishable academic paper, thereby addressing the ambition of making policy and decision-making more accessible, by officials disseminating their own accounts of policy and practice in Cape Town. By engaging with theory and international case studies and best practices, the ambition of the program is for there to be a positive feedback loop into the City after the exchange period to improve the defensibility of policy and decisions.

From CityLabs to KTP: methodological reflections

Most of the CityLabs involved interaction between academic researchers (mainly from UCT) from a range of disciplines and practitioners (from government and elsewhere) from a range of sectors, and generally started with one or two open-ended discussions followed by a series of seminars on a range of issues relating to the topic of each particular Citylab. The CityLabs generally culminate in a publication that includes chapters/articles written by both academics and practitioners. In some cases, collaborative research projects were also identified. In the case of the Climate Change CityLab, there was a formal partnership (the Climate Change Think Tank) between the City, ACC, and other organizations to co-produce climate change-related research on Cape Town to feed into the City's strategies. Diverging methods used by different researchers can in part be attributed to the different topics and states of knowledge in the field. However, it also represents the different ways of understanding what co-production is and should be among the different CityLab coordinators.

Some of the seminars had less participation than hoped for from government, which was a result of most City and Province officials being overburdened with work and some of the topics not being of immediate relevance to their work. In addition, time constraints meant that officials were often unable to play a significant role in the collaborative research projects. The CityLab program and the KTP have each tried to address these constraints in various ways. One means through which the KTP attempts to overcome the capacity constraints of City officials is to engage in co-production processes by embedding researchers in the City administration for three years. The motivation for this is twofold: to enable the researchers themselves, at least in part, to straddle the worlds of academic research and local government practice; and to free up time for City officials to engage in the co-production processes.

During 2012, ACC and the City co-hosted a workshop on co-production to learn from the experiences of the CityLabs and thereby inform the approach in the KTP. All of the CityLab researchers as well as City officials and the KTP researchers attended this workshop. In an attempt to establish a baseline with regard to expectations and understandings of co-production among participants, participants were divided into mixed groups tasked with graphically representing their understanding of co-production. This process was revealing, as it highlighted that even among the CityLab and KTP researchers and their City counterparts, there is no single expectation about and approach to co-production.

Reflections on normativity and co-production at the ACC

The ACC's CityLab program and KTP have attempted to engage with the challenges of normativity in various ways. In some cases, these processes have been explicit while in other cases they have been implicit. In this section, we ask specific questions which draw attention to the normativity of specific projects. We draw on examples from individual CityLabs and the KTP to highlight the normative challenges and how they were negotiated, sometimes with and sometimes without explicit consideration of the norms that informed decisions. This is intended to demonstrate through examples the importance of normativity in knowledge co-production, and the need to give more explicit attention to the normative choices made during the co-production process. Our reflections are structured around four key issues: defining the research question or problem; identifying the intended impacts; choosing research methods; and identifying who is involved and how.

Defining the research question and/or problem

In most cases, the CityLabs were initially very open-ended, with only a broad topic to structure participants' engagement, such as urban flooding in Cape

Town. Although ACC initiated most of the CityLabs from the starting point of its normative vision and mission, participants in the CityLabs collaboratively defined the problem and over time developed a normative position. For example, the Philippi CityLab initially had no specific research agenda, but after about a year of meetings with interested organizations and individuals an agenda was developed. In the case of the Healthy Cities CityLab, a particular, fairly narrow approach was also adopted early on in the process, but was subsequently made much more open-ended through a long process of engagement among researchers interested in taking part. An initial focus on the relationship between the built environment and chronic disease at the neighborhood scale was established because that was where the research interests of some of the key participants at the first CityLab meetings overlapped. Through a long process of extensive engagement (and frequent disagreement) among participants in the CityLab, the majority of participants decided that the focus should be broadened as the research question needed to be more holistic, and that a wider range of health and wellbeing issues at various scales from the household to the city as a whole needed to be looked at. It was also decided that there needed to be engagement with the City around ways in which they could create healthier urban environments.

In the case of the Urban Flooding CityLab, although there was an open-ended series of seminars, obtaining funding for a collaborative research project meant that certain choices had to be made at an early stage in the process, which resulted in a particular theoretical lens (collaborative governance) being adopted for the research project.

The experiences of the CityLabs highlight the importance of collectively defining the problem and framing the research question. This is not to suggest that particular framings are problematic, but instead that frames carry implicit narratives regarding the relationship between actors, problems, and solutions. Ideally this needs to be done collectively by researchers and practitioners involved in the co-production process, and sufficient time should be allowed for understanding alternative approaches. It should be recognized that this will often be a messy process with conflicting interests, but in all cases the CityLabs were able to formulate problems and approaches that allowed for participation by a range of disciplines and sectors.

Identifying the intended impacts

In many cases the CityLabs were not started with specific impacts in mind. Instead, the aim was to explore issues and expose policymakers and practitioners to a broader range of debates and perspectives, thus stimulating a more holistic view of the problem and potential solutions. For example, the Climate Change CityLab brought together academic researchers and practitioners from different sectors to engage with and provide input into the City's research and policy work on climate change adaptation and mitigation. The operations of the Climate Change CityLab were noted to have helped to drive a "shift in how City officials view climate change, taking it from what was previously perceived as exclusively an environmental issue to one of diverse relevance" (Anderson *et al.* 2013: 6).

In some cases the CityLabs had a specific set of impacts in mind from the outset, driven by a particular normative viewpoint. In the case of the Central City CityLab, one of the key aims was to encourage urban densification. Although densification is a contested objective, resisted by many politicians and residents, the promotion of densification is based on the long-standing view of progressive urban planners that this will create more equitable and sustainable cities (Todes 2006). This normative perspective is held by most of the urban planning staff of the City of Cape Town, and is captured in all of the City's spatial planning strategy documents. In collaboration with the Cape Town Partnership (CTP), the City developed the Central City Development Strategy, which proposes to increase the Cape Town central city population by 100,000 people within a ten-year period. As a starting point, the Central CityLab attempted to understand the political, institutional, and practical obstacles to the densification of central Cape Town. The Central City CityLab's clear normative agenda attracted participants from different sectors and disciplines who supported the goal of densification and who came together in a relatively coherent set of seminars and to produce a coherent draft book. It is important to note that being explicit about the normative goals of the CityLab did not prevent debate and critical views. Some of the work undertaken for the Central City CityLab challenged the normative goal of densification, highlighting contradictions and wrong assumptions in strategies to promote densification (for example, Turok 2011), which greatly contributed to the richness of the discussions in, and outputs from, the CityLab.

The experience of the Central City CityLab shows the benefits of having participants in knowledge co-production processes be explicit about the changes they want to see in the world, and about the assumptions and theories underpinning this. For groups like the CityLabs, there may only be a small amount of overlap in what different participants want to change in the world in relation to that particular theme, but it is important to identify and be explicit about this (while not excluding critical voices).

Choosing research methods

Closely linked to defining the research problem and identifying desired impacts is the choice of research methods and tools. In most cases, CityLabs embraced a wide range of (sometimes conflicting) methods and tools, informed by the diversity of participants. In some cases, however, new collaborative interdisciplinary research projects were initiated, and specific research methods and tools needed to be chosen.

In the case of the Healthy Cities CityLab, reaching agreement on research methods and tools was particularly difficult, as the CityLab included a wide range of disciplines, from the health sciences to social sciences. In general, most researchers from the health sciences have a positivist approach, relying on large-scale quantitative surveys, standardized measuring instruments, and statistical analysis. Many social scientists, on the other hand, see qualitative research methodologies, such as open-ended interviews and focus groups, as being important. An attempt was therefore made to combine these quantitative and qualitative methodologies. Initially the intention was to use these methodologies in parallel. After discussing the matter, however, participants in the CityLab agreed that many of the usual quantitative methodologies and tools were inappropriate to our context, and that the qualitative research component therefore needed to precede the quantitative research, so that it could inform the standardized quantitative research methodologies and instruments and adapt them to our local context.

This example highlights the need for extensive engagement about the choice of research methods and tools and the assumptions and perspectives underpinning them. The experience from the Healthy Cities CityLab suggests that mixed research methods and a variety of research tools are usually necessary for interdisciplinary research (see also Leonelli 2005).

In the KTP, the embedding of Ph.D. researchers in the City administration for seven months of the year over a three-year period provides an opportunity for the researchers to gain access to City processes of policy development and formulation, high-level discussions, and decision-making processes—aspects of local government not usually accessible to academic researchers. The City has a wealth of data that is not usually made available beyond the organization, and the embedded researchers' access to this has sometimes made it possible to package the data for alternative uses. For example, the Economic Areas Management Program (ECAMP) data is being made available throughout the City, and will eventually become accessible to anyone through the City of Cape Town's website.

Who is involved and how?

The persons and organizations that are involved in knowledge co-production processes and the extent and nature of their involvement have a profound impact on all the above issues—defining the research problem, identifying the impacts to be achieved, and choosing research methods and tools. The CityLabs had a broad range of participants, including academics from various disciplines, officials from the City and Province, community organizations, NGOs, students, and the private sector. In addition, the participants in the various CityLabs were involved in different ways. For instance, they could be seminar presenters, seminar attendees, and/or members of working groups. Each CityLab has a slightly different means of enrolling participants, and these different methods entail different degrees of control over who participates. In most cases, seminar invitations were sent out to everyone on ACC's large mailing list, and anybody interested could attend. In some cases, such as the Healthy Cities CityLab, participants potentially interested in the interface between health and the built environment were identified in advance, and invited to initial meetings, after which additional invitees were identified. In all the CityLabs, the working groups that undertook collaborative research projects and the writing of publications were generally small groups constituted largely through self-selection. More specifically, they were made up of people who wanted to be involved and who were able to dedicate time to the process and attend meetings.

The experience of the CityLabs shows that there is a need to understand and be explicit about what kind of knowledge community is desired, and to choose the methods for involving participants accordingly. For example, the Urban Ecology CityLab used a seminar series, housed at UCT, which resulted in the building of a community of experts, both inside and outside of the university, although the attendees typically differed for different topics (Anderson *et al.* 2013). In the Philippi CityLab, in contrast, academic participants were interested in a seminar series, whereas the community and NGO participants sought more direct involvement in development-related activities (Brown-Luthango 2013). In an attempt to facilitate involvement by communities and NGOs, the seminars alternated between UCT and various venues in Philippi. The different composition of the core members of the two CityLabs resulted in very different second phases. In the Urban Ecology CityLab, the second phase entailed the creation of a research group with an emphasis on experimental work to build an empirical base to elucidate the urban-specific nature of the ecology of Cape Town (Anderson and Elmqvist 2012). In contrast, in the Philippi CityLab, a framework was adopted in the second year to guide the engagement, and focus was placed on "integrated community development." This second year adopted an approach much more directed at setting up partnerships among those with key roles, in particular between the CityLab and NGOs in Philippi. In both cases, the second year had much greater continuity in participation.

The experiences of the Urban Ecology and Philippi CityLabs highlight the need to give explicit thought to who is invited, how they are invited, and where and how meetings take place. Proactive action may be required to ensure that a sufficiently wide range of disciplines and sectors are given the opportunity to be involved. In addition, in the CityLab program it was found that individual researchers' background, experiences, values, and ideologies influenced who ended up participating in each CityLab (Anderson *et al.* 2013).

Mapping out these networks and values is also important for the Knowledge Transfer Program. In this case, the City administration selected four researchers to be embedded in the city. However, these individuals were not selected randomly, or even based on their academic skills. Instead, they were chosen because they were experienced researchers with established relationships to City officials, and who were capable of acting as intermediaries, with a commitment to the vision of both the ACC and the City. In the KTP, it is understood that the embedded researchers have brought different knowledges to the City, sometimes with a more theoretical underpinning. It has been noted that this knowledge is helping to set a clearer direction for policy and strategy development in the City in the areas in which the embedded researchers are working. The City has also benefited from moving out of a traditional, consultancy-type relationship with the university on these particular topics. (Consultancy relationships are typically characterized by a one-directional flow of information and work.) The KTP attempts to address this power imbalance without conflating the roles of officials and academics. The flow of knowledge is thus now more bi-directional, feeding into research agendas as well as policy processes to strengthen both. Similarly, the participants in the program from the City are inevitably those who see value in the academic partnership. At a strategic level, the Project Steering Committee and operational Project Advisory Committee govern the KTP. Both these committees contain equal numbers of members from the City and the ACC, and both are co-chaired by members of each of the partner institutions. The shared governance of the program is intended to encourage shared and equal ownership of the program and its outcomes.

Implications for knowledge co-production

Being explicit about normativity is important for *all* research, however, as we have attempted to highlight in this chapter, it is particularly important where participants from multiple disciplines and sectors engage in co-production processes intended to have an impact on policy and practice. A key learning from the CityLab program is that it is important to work with stakeholders to develop an explicit set of shared norms over time. Real co-production ideally needs to involve all partners at all stages of the process (ideally from the problem identification stage onwards), which implies a very open-ended and exploratory process at the beginning. Most of the CityLabs followed a sequence like this, initially starting with just an open-ended topic, and then, through a process of engagement over time, ending up with a more defined agenda. The KTP was different in that detailed topics were identified beforehand by the ACC and City, but doing so required extended negotiations. It is important that the normative agenda, once decided upon, be as explicit as possible. However, being explicit about the normative agenda, while at the same time ensuring that participants from a range of disciplines and sectors participate in the co-production process, is not an easy balancing act.

Four further points emerge from our reflections on normativity and co-production. The first is that *different participants and institutions have overlapping yet unsynchronized working processes.* This perspective recognizes that the objectives, timelines, and expected outputs of different institutions vary. Government institutions such as the City generally have sporadic periods of intense demands and pressing needs that must be responded to urgently, whereas, by contrast, the work patterns in academia are usually tied to the individual researchers' rhythms of academic work. Government officials are also driven by more concrete outputs, and are much less invested in, and rewarded for academic publication. At the intersection of erratic government processes and more controlled research processes is the process of co-production, which must respond to the needs of all the participating institutions. There is a need for "intentional co-creation," where there is a practical alignment based on a commonality of normative objectives shared between institutions. Finding points of intersection in the rhythms of work is also important to successfully undertaking co-production.

The second point we draw attention to is that co-production *provides participants with a glimpse of doing things differently*. Here, the focus is on innovation and the introduction of new norms, practices, and configurations. There is an acknowledgment that this process is necessarily deliberative and even conflictual. For example, while reducing flooding in informal settlements is a relatively vague objective that most stakeholders would agree with, the views on how best to understand and address flooding, and the roles and responsibilities of various stakeholders in this, may vary enormously. The CityLabs provided an opportunity for participants to engage with others who view problems, causes, and solutions differently—not just to read their perspectives in text, but to put faces and personalities behind the stories. This tension leads to deeper engagement to find solutions that fit all institutions involved in the process of co-production. Trust, feedback, and reflexivity, shaped by a common vision of maximizing the public good, are regarded as key ingredients in co-production.

The third point is a question regarding whether the *outcome of knowledge co-production is knowledge or learning*. In engaging different stakeholders, one of the key outcomes was greater understanding across different sectors and disciplines. This, however, did not necessarily lead to new knowledge in the traditional academic sense of publishable research findings. The most common outcome was that individuals sought ways to incorporate the new ways of understanding into their established views. The methodological process thus became the focus; ways of integrating knowledge that is new to individuals became more important than creating "new" knowledge.

Finally, our experiences in co-producing knowledge generated important insights into *the limitations of knowledge itself*. Often in CityLab and KTP discussions, participants raised the issue of vested interests, powerful entities, and established, difficult-to-change patterns of accumulation. In these cases, knowledge provided useful insights, but did not make it possible to effect change. This challenges one of the implicit assumptions embedded in the co-production model: that better knowledge can and will be adopted, and will lead to change. Instead, this insight highlights that knowledge is not the only factor that creates change: understanding power is also a critical part of negotiating co-produced solutions, and in many cases knowledge alone is insufficient to implement or instigate change.

Although the focus here has been on University–City relationships, there are of course no restrictions on the types of institutions involved in co-producing responses to complicated urban transitions. Similarly, the different representations of co-production indicate that the objective of a co-production relationship can vary from sharing knowledge and understanding problems from one another's perspectives (and building capacity to engage/participate), to influencing processes by generating new understandings and articulations of urban problems and, in turn, influencing responses and outcomes. While there is no set approach to co-production in Cape Town, there is nonetheless a recognition that these processes are highly political, and power is constantly redistributed between parties.

In conclusion, we suggest that recognition of the situatedness of research requires us to be more careful and specific about the motivations for particular decisions, and, specifically, to ensure that the normative agendas which motivate and inform decision-making are clear. We suggest that doing so will help improve the development of theory on the co-production of knowledge, enabling better comparisons, explanations, and generalizations.

Notes

1 The DA has governed the City of Cape Town since March 2006 and the Western Cape Province since May 2009.

2 The degree of genuine ideological difference between the DA and ANC remains highly contested. McDonald, for example, suggests that "the *general* character of post-apartheid ANC urban policy-making is solidly neoliberal in its orientation" (McDonald 2008: 69), whereas others disagree (for example, Parnell and Robinson, 2012).

3 University of Cape Town, Stellenbosch University, University of the Western Cape, Cape Peninsula University of Technology. URAP (2013) ranks the first three of these as first, third, and fifteenth in Africa respectively in terms of academic performance.

References

Anderson, P. M. L., Brown-Luthango, M., Cartwright, A., Farouk, I. and Smit, W. (2013) 'Brokering communities of knowledge and practice: Reflections on the African Centre for Cities' CityLab programme', *Cities*, 32: 1–10.

Anderson, P. M. L. and Elmqvist, T. (2012) 'Urban ecological and social–ecological research in the city of Cape Town: Insights emerging from an urban ecology CityLab', *Ecology and Society*, 17(4): 23 [online]. Available at: http://dx.doi.org/10.5751/ES-05076-170423 [accessed June 16, 2014].

Boersma, F. K., Reinecke, C. J. and Gibbons, M. (2008) 'Organizing the university–industry relationship: A case study of research policy and curriculum restructuring at the North–West University in South Africa', *Tertiary Education and Management*, 14(3): 209–26.

Brown-Luthango, M. (2013) 'Community–university engagement: The Philippi CityLab in Cape Town and the challenge of collaboration across boundaries', *Higher Education*, 65(3): 309–24.

Bulkeley, H. (2006) 'Urban sustainability: Learning from best practice?', *Environment and Planning A*, 38: 1029–44.

Carolan, M. S. (2007) 'Sustainable agriculture, science and the co-production of "expert" knowledge: The value of interactional expertise', *Local Environment*, 11(4): 421–31.

City of Cape Town (2011) 'Urbanisation and poverty in Cape Town'. Working document. Cape Town: City of Cape Town.

——(2012) *City of Cape Town – 2011 Census – Cape Town*. Available at: www.capetown.gov.za/en/stats/Documents/2011%20Census/2011_Census_Cape_Town_Profile.pdf [accessed June 16, 2014].

Crane, W. and Swilling, M. (2008) 'Environment, sustainable resource use and the Cape Town functional region – An overview', *Urban Forum*, 19(3): 263–87.

Ernstson, H., Lawhon, M. and Duminy, J. (2014) 'Conceptual vectors of African urbanism: "Engaged theory-making" and "platforms of engagement"', *Regional Studies* [online]. Available at: http://dx.doi.org/10.1080/00343404.2014.892573 [accessed June 16, 2014].

Flyvbjerg, B. (2004) 'Phronetic planning research: Theoretical and methodological reflections', *Planning Theory and Practice*, 5(3): 283–306.

Goebel, A., Hill, T., Fincham, R. and Lawhon, M. (2010) 'Transdisciplinarity in urban South Africa', *Futures*, 42(5): 475–83.

Hammett, D. (2012) 'W(h)ither South African human geography?', *Geoforum*, 43(5): 937–47.

Hessels, L. K. and van Lente, H. (2008) 'Re-thinking new knowledge production: A literature review and a research agenda', *Research Policy*, 37(4): 740–60.

Jansen, J. D. (2002) 'Mode 2 knowledge and institutional life: Taking Gibbons on a walk through a South African University', *Higher Education*, 43(4): 507–21.

Jasanoff, S. (ed.) (2004) *States of knowledge: The Co-production of Science and Social Order*. Oxford: Routledge.

Klein, J. T. (2004) 'Prospects for transdisciplinarity', *Futures*, 36(4): 515–26.

Lawhon, M. and Patel, Z. (2013) 'Scalar politics and local sustainability: Rethinking governance and justice in an era of political and environmental change', *Environment and Planning C*, 31(6): 1048–62.

Lemanski, C. (2007) 'Global cities in the South: Deepening social and spatial polarisation in Cape Town', *Cities*, 24(6): 448–61.

Lemos, M. C. and Morehouse, B. J. (2005) 'The co-production of science and policy in integrated climate assessments', *Global Environmental Change*, 15(1): 57–68.

Leonelli, S. (2005) 'Pluralism and normativity in interdisciplinary research', *Graduate Journal of Social Science*, 2(1): 1–6.

McDonald, D. A. (2008) *World City Syndrome: Neoliberalism and Inequality in Cape Town*. New York: Routledge.

Maharaj, B. and Narsiah, S. (2002) 'From apartheid apologism to post-apartheid neoliberalism: Paradigm shifts in South African urban geography', *South African Geographical Journal*, 84(1): 88–97.

Mitlin, D. (2008) 'With and beyond the state – Co-production as a route to political influence, power and transformation for grassroots organizations', *Environment and Urbanization*, 20(2): 339–60.

Oldfield, S., Parnell, S. and Mabin, A. (2004) 'Engagement and reconstruction in critical research: Negotiating urban practice, policy and theory in South Africa', *Social & Cultural Geography*, 5(2): 285–99.

Parnell, S. (2007) 'The academic-policy interface in post-apartheid urban research: Personal reflections', *South African Geographical Journal*, 89(2): 111–20.

Parnell, S., Pieterse, E. and Watson, V. (2009) 'Planning for cities in the global South: An African research agenda for sustainable human settlements', *Progress in Planning*, 72(4): 233–41.

Parnell, S. and Robinson, J. (2012) '(Re)theorizing cities from the global South: Looking beyond neoliberalism', *Urban Geography*, 33(4): 593–617.

Pestre, D. (2003) 'Regimes of knowledge production in society: Towards a more political and social reading', *Minerva*, 41(3): 245–61.

Petts, J., Owens, S. and Bulkeley, H. (2008) 'Crossing boundaries: Interdisciplinarity in the context of urban environments', *Geoforum*, 39(2): 593–601.

Pickersgill, M. (2012) 'The co-production of science, ethics, and emotion', *Science, Technology & Human Values*, 37(6): 579–603.

Pieterse, E. (2013) 'City/university interplays amidst complexity', *Territorio*, 66: 26–32.

Pohl, C. (2008) 'From science to policy through transdisciplinary research', *Environmental Science & Policy*, 11(1): 46–53.

Pohl, C., Rist, S., Zimmermann, A., Fry, P., Gurung, G. S., Schneider, F., Speranza, C. I., Kiteme, B., Boillat, S., Serrano, E., Hadorn, G. H. and Wiesmann, U. (2010) 'Researchers' roles in knowledge co-production: Experience from sustainability research in Kenya, Switzerland, Bolivia and Nepal', *Science and Public Policy*, 37(4): 267–81.

St Clair, A. L. (2006) 'Global poverty: The co-production of knowledge and politics', *Global Social Policy*, 6(1): 57–77.

Sanders, M. (2002) *Complicities: The Intellectual and Apartheid*. Durham, NC: Duke University Press.

Schmidt, D. (2008) 'From spheres to tiers: Conceptions of local government in South Africa in the period 1994–2006', in M. van Donk, M. Swilling, E. Pieterse and S. Parnell (eds), *Consolidating developmental local government: Lessons from the South African experience*. Cape Town: UCT Press, pp. 109–30.

Shinn, T. (2002) 'The triple helix and new production of knowledge: Prepackaged thinking on science and technology', *Social Studies of Science*, 32(4): 599–614.

Swilling, M. (2010) *Sustaining Cape Town: Imagining a Livable City*. Stellenbosch: Sun Press.

Todes, A. (2006) 'Urban spatial policy', in U. Pillay, R. Tomlinson and J. du Toit (eds), *Democracy and delivery: Urban policy in South Africa*. Cape Town: HSRC Press.

Turok, I. (2001) 'Persistent polarisation post-apartheid? Progress towards urban integration in Cape Town', *Urban Studies*, 38(13): 2349–77.

——(2011) 'Deconstructing density: strategic dilemmas facing the post-apartheid city', *Cities*, 28(5): 470–7.

Tuanneinen, J. (2005) 'Hybrid practices? Contributions to the debate on the mutation of science and university', *Higher Education*, 50(2): 275–98.

United Nations Human Settlements Programme (UN-HABITAT) (2010). *State of the World's Cities 2010/2011: Bridging the Urban Divide*. London: Earthscan.

University Ranking by Academic Performance (URAP) (2013) '2010–11 ranking by region: Africa'. Available at: www.urapcenter.org/2011/region.php?rcode=AF [accessed June 16, 2014].

Wilkinson, P. (2000) 'City profile: Cape Town', *Cities*, 17(3): 195–205.

——(2004) 'Renegotiating local governance in a post-apartheid city: The case of Cape Town', *Urban Forum*, 15(3): 213–30.

4 Tracing Kisumu's path in the co-production of knowledge for urban development

George Mark Onyango and Bernard Otieno Obera

Introduction

Kisumu city has over the years positioned itself as a unique urban area, one where interventions have been piloted and knowledge co-produced and shared across the East African divide. Co-production of knowledge has been closely linked to urban development in this lakeside city. The structure of urban systems in Kenya has been linked to knowledge innovation, be it either imported or locally produced. We therefore note that a historical analysis provides the best framework for appreciating the concept of co-production in urban systems.

Kisumu, the third largest urban area in Kenya, is strategically located—at the tail end of the railway line from Mombasa, on the shores of Lake Victoria, and near the borders with Uganda and Tanzania—and so acts as a gateway to the Great Lakes Region. As the city has grown, it has attracted a number of institutions including universities and research organizations. This amalgam has created a ripe environment for the co-production of knowledge critical for sustainable urban development. Co-production in this context includes processes, experiences, and interventions that have involved different players who have worked together and seen the knowledge they produce be used as a basis for influencing the city's development.

In this chapter we trace the history of Kisumu from 1900 to the present to demonstrate the various levels of knowledge co-production over the years and how these have shaped the development of the city. In certain instances the process might not be so evident, but viewing the events cumulatively one begins to discern the knowledge production taking place over time.

The historical basis for knowledge co-production in Kisumu

Colonial period, 1900 to 1950

Kisumu is said to have been a versatile commercial center before the advent of colonialism in the late nineteenth century. The native Luo community traded with the neighboring Luhyias, Kalenjins (especially the Nandi sub tribe), and Kisii ethnic communities. In interviews with senior citizens knowledgeable about Kisumu's history, it has been suggested that the shores of the lake offered open ground with easy access and the possibility to detect ambushes by enemies. Such ambushes were common, especially from the pastoral Nandi community, well known for cattle rustling. The location thus developed as a market place ("Kisuma") guided by the realization that the site was neutral ground, allowing for peaceful trade. All the three ethnic communities (Luos, Kalenjins (Nandi sub tribe), and Kisiis), in one way or another contributed toward the peaceful operation of this market center (Onyango and Ogonda 1999).

Lonsdale (1989) observes that in the ten years between 1895 and 1905 "Kenya," if such an anachronistic concept may be permitted, was transformed from a 600-mile-long footpath into a colonial administration. This transformation was the work of force. This conquest of the peoples of Eastern Africa disrupted the trading strategies that had linked the peoples of Nyanza basin who had already reached a working equilibrium.

The Uganda railway introduced new peoples, cultures, goods, and services. The struggle in the British parliament between the people for and against the Uganda railway was summed up in a verse in a British newspaper as follows:

> What it will cost no words can express
> What is its object no brain can suppose
> Where it will start from no one can guess
> Where it is going nobody knows
> What is the use of it none can conjecture
> What it will carry there's none to define
> And in spite of George Curzon's superior lecture
> It clearly is naught but a *Lunatic line.*
>
> (Miller 1971: 286)

But Curzon, the then under-secretary of State for Foreign Affairs, won the day. He convinced Parliament that the Protectorate of Uganda was absurd without a railway to the coast (ibid.). The proposed railway would start from Mombasa, run parallel to the caravan trail established earlier by the Imperial British East Africa Company, cross the Rift Valley at Elgeyo ranges, and run parallel to the Nzoia River down to the regional capital of Port Victoria.

Colonial development was driven by colonial interest with little opportunity for mutual agreement on the direction of this development. The local community was relegated to a passive role when it comes to contributing knowledge to the town's development (Anyumba 1995). The Asian community was introduced into the town's socio-economic fabric, based on their longer history of relations with the British and their technical expertise as railway technicians and traders. A small portion of the British citizens also moved to Kisumu and were soon followed by other ethnic groups from Kenya like the Teso and Somalis to offer some services locally considered taboo, like the "night soil" services in the absence of a sewer network.

In 1908 an outbreak of bubonic plague led to the zoning of residential areas in the town. Zone A included the port, official residences for colonial officers, the government and railway headquarters, the jail, police housing, and the hospital, and was occupied by Indians and Europeans employed in the town. Zone B was a buffer between Zones A and C. Zone C was the official African residential area. However beyond Zone C a new settlement of African villages was arising just outside the town boundary. This area was generally poorly developed with semi-permanent housing structures and very poor sanitation facilities.

In 1930 the town boundary was reduced "to make it more manageable." This reduction in the size of the town effectively excluded the settlement areas, beyond Zone C, from the town. The residential area therefore developed as a peri-urban unplanned settlement, with no basic infrastructure (Obudho and Waller 1976). On the other hand the areas in Zone A, which were provided with infrastructure, grew in an organized manner, and housed the better-off urban residents. After independence in 1963, Zone B was slowly invaded by Zone A as the demand for land within the central parts of the town grew. Eventually what was formerly Zone B was absorbed by the expanding Zone A, and it is now part of the inner city. The net effect of this zonal segregation was a skewed urban development, guided by the type of knowledge prevalent among the dominant community in each zone. The peri-urban settlements developed in accordance with indigenous knowledge about settlements, with the typical Luo homestead being the preferred layout. A few people returning from Asia after the Second World War introduced rooming/row housing, borrowed mainly from the Indian peninsula; and European urban-planning ideas dominated the housing types and settlement patterns in Zone A.

Colonial period, 1950 to 1963

The period before independence saw the return of many people from service in the Second World War. Having interacted with other people from the rest of Africa and the world, they were more conscious of their potential role as democratically involved citizens. Hence they began agitating for inclusive governance and better living conditions. This developed into a national uprising, which led to the imposing of more stringent policies to control the local communities and force compliance. The colonial government declared a state of emergency in 1952, when the native Kenyans violently rose up against discrimination and their exclusion from the running of their motherland and the management of her resources. The Mau Mau movement was to shape the nation's future development. The Kisumu Municipal by-laws were introduced in 1954 to further coerce the indigenous people into adopting western culture; the African culture and knowledge base were considered archaic. Institutions of higher learning in Europe also became centers of agitation for equality toward the end of this period. This was due to African students from various countries meeting each other, sharing their stories, and linking them to what they were learning. Students from Kisumu and the Nyanza region were able to take advantage of the less violent resistance in the lake region to get formal education (Onyango 1998). This later contributed to the region's easier acceptance of external partners in knowledge co-production, making Kisumu a popular city for external players involved with urban development.

The post-independence period, 1963 to 2000

Kisumu was declared a municipality in 1960, on the eve of independence, with the first African mayor (Cllr. Alderman Mathews Ondiek) being elected two years later, followed by the first woman mayor (Cllr. Grace Onyango). Kenya's independence, while celebrated as marking a new dawn, turned out to be a pipe dream. The hopes of transferring the management of indigenous resources into local hands, with improved quality of living and free services as a result, were shattered. The educated elites and their cronies took over the role of the colonial administrator, setting the stage for a future government where awards such as appointments were based on clan affiliations (Sandbrook 1975). This was to be a cause of major friction between the leadership in Kisumu (and indeed Luo-dominated Nyanza province) and the new government. The then Kenyan Vice-President, Jaramogi Oginga Odinga, who hailed from Nyanza, was later to resign in protest due to the level of corruption of the independence government. The assassination on July 5, 1969 of the youthful Minister for Justice Tom Mboya, who hailed from Nyanza, was a major turning point in the power relations between the government and the people of Kisumu. This was demonstrated by the public confrontation between Mr. Oginga and President Jomo Kenyatta on October 25, 1969 at the opening of the New Nyanza Provincial General Hospital, in Kisumu. The two incidents were to spark violent riots during which the police in Kisumu and Nairobi killed scores of people. This caused a significant rift between the central government and the Kisumu people and marked the birth of opposition politics in Kisumu, leading to the town's marginalization in national development interventions. The one-party system that emerged in Kenya after these two incidents shaped the future of Kisumu's development, which was modeled on a uni-centric system in which the central government implemented its national agenda without regard for local knowledge.

During this period, local Luo clan rivalry emerged in Kisumu. Kisumu's three large clans (Kisumu, Kano, and Kajulu) joined forces to oppose the Nyakach clan's apparent dominance in the management of the city resources. This rivalry was born of the fact that the first African town clerk of Kisumu, Mr. Ojwang K'Ombudo, employed many people from his Nyakach clan as well as enabling them to acquire land in the town. Economic conditions were becoming harsher, with failing industries and a rise in the informal sector, and the rivalry manifested itself in competition over the limited space on the streets and in open areas for informal trade, as well as in the struggle against the local authorities that were believed to be dominated by Nyakach clans. The informal sector was a constant headache for the Municipal Council. It kept growing, with groups setting up business in almost any available space, leading to constant street battles with the Municipal Council enforcement officers. From interviews with two reformed members of the vigilante group Bagdad Boys, Audi Ogada and Seth Ochieng, it has emerged that the local politicians took advantage of this situation to agitate for change and their own selfish aggrandizement. For instance Cllr. Lawrence Oile, mayor of Kisumu in the 1990s, is said to have propagated this dreaded youth group (dubbed the "Bagdad Boys" in the wake of the American invasion of Iraq in the 1990s) in the informal sector. Backed by the government party, KANU, Cllr. Oile deployed the group to try to suppress the agitation for political change, and fight off Raila Odinga and the change he represented.

The conflicts between the informal sector and the Municipal Council were to become a serious issue in the 1990s as the clamor for multiparty democracy took shape. The town became largely ungovernable with failing basic urban services including water, roads, and housing among others. Attempts at dialogue met with no success, and the slightest unrest would blow up and affect the whole town, giving Kisumu the unenviable status of the most violent town in Kenya during the struggle for multiparty democracy. This label proved difficult to shed as the situation was further aggravated by the poor state of the basic services. The water and sewage services were never expanded, no new public housing was being constructed, and unemployment continued to rise. The need for an injection of external ideas and resources became apparent. The indigenous community in and outside Kisumu, especially in Nairobi, teamed up with local actors to start a dialogue for change, giving birth to the "Kisumu we want" campaign, in recognition of the residents' desire for a better town (Ufadhili Trust 1997). This was the first properly documented process of knowledge co-production involving the citizens, the city managers, practitioners, and academia, led by civil society organizations who took the lead in this new way of working together. An iterative process was established with facilitation by academia in which various issues were discussed and ideas were developed, consolidated, and documented in the report "Kisumu We Want" (ibid.). It enabled the people of Kisumu to appreciate that they could all be partners in the process of producing new ideas to facilitate their town's development. The document was tabled in the Municipal Council for adoption and translation into a policy that would guide urban development. Despite this promising beginning, the Municipal Council of Kisumu found it difficult to understand the report and was reluctant to buy in. Due to this lack of government support, the initiative never reached very far in terms of implementation, however it did form the basis for an appreciation of knowledge co-production and ultimately awakened the various stakeholders to their capacity to bring about change. A number of organizations moved into Kisumu to build on this initiative, probably realizing that Kisumu offered a different model of citizen engagement in Kenya.

The town soon gained the reputation of having the largest number of NGOs per square kilometer! A number of these NGOs adopted the model of including citizens, the city managers, practitioners, and academia in their information profiling processes.

The Scandinavian countries were the first to formally raise concerns about the state of the environment of the cities around Lake Victoria and the relationship between the lake and people's livelihoods. At the Kyoto meeting on the environment in 1997, the mayors of Kisumu, Mwanza, and Entebbe met officials from cities around the Baltic Sea who faced similar issues concerning the status of shared water bodies. The Baltic officials helped the mayors form the Lake Victoria Region Local Authorities Cooperation (LVRLAC), based on the Union of the Baltic cities model (LVRLAC 2010). This model enhanced the possibilities for mutual learning and for sharing experiences and resources to address local challenges, as well as to create knowledge co-production and exchange from a North–South global perspective. Concerned about the growing poverty and its adverse effects on Lake Victoria, the Swedish International Development Agency (Sida) commissioned a study on the environmental situation of the cities around Lake Victoria (Carl Bro Consultants 2001). This raised issues touching on the sustainability of the cities and their environment. The study was intended to inform Sida's commitment to the Lake Victoria Region (Seman 2003). Other programs were also initiated locally with the support of the Finnish and Dutch governments. During this stage the nongovernmental organizations (NGOs) began in earnest to align themselves with the new donor interests and to link up with local opportunities for community empowerment. This was probably also facilitated by the government having become more tolerant of the dissent associated with NGOs striving for civic empowerment of the communities.

The late 1990s saw the emergence of a strengthened civil society sector in Kenya through the emergence of civil society organizations (CSOs), which had ramifications in Kisumu. This caused some ambivalence, however, with the urban authorities seeing the CSOs as a menace, and the locals, especially the informal sector, treasuring them as a team member in the fight against oppression by the authorities. This period also saw the entrance of academia to fill the gap in co-production of the knowledge required to guide the programming of the CSOs. In developing intervention strategies for urban development, these programs required a combination of professionals and buy-in of citizens and implementing agencies, including the Municipal Council. Academics were viewed as objective actors in this process. Seeing the research opportunities that were opening up, some academics were quick to craft consultancy outfits operating as private enterprises to offer these services (Mitulla 2004). The CSO boom soon led to fatigue among the local community, due to the multiple consultations by the CSOs and the knowledge co-production enterprises operating as consultancy firms. The CSOs were competing for the same resources from a narrow group of donors as well as for the hearts and souls of the same community. There were cases of CSOs claiming credit for each other's work, with some conducting parallel activities, even in the same sites. A need for coordination was noted by the various implementing agencies. To fill this need, the Western Kenya NGO Network was born, supported by Action Aid. However, this initiative was short lived and the confusion, conflict, and duplication went on. The private sector also had conflicts, both internally and with the informal sector. The Municipal Council was accused of collecting revenue without providing services, while some corrupt officials colluded with members of the private sector to grab public land, displacing the members of the informal sector. One example is when the old municipal slaughterhouse located on public land at the lakeshore near the railroad yard was fenced off by a business group forcing the informal businesses running food kiosks, known as Lwangni hotels to move to their current site near the lake (Proceedings of the first stakeholder meeting 2008; UN-Habitat 2005a, 2005b).

The new millennium: civic engagement in the development of Kisumu after 2000

The continuous strife with and among the street traders was increasingly understood as a manifestation of citizens' frustration with the poor governance and the accompanying poor service provision of the local authorities. Members of the business community were also increasingly showing their frustration with the lack of basic services for which they had paid, as well as with unfair competition from the informal sector that encroached on the shop/factory doorsteps. Various private sector groupings emerged in response to these threats from the informal sector. The Kenya Association of Manufacturers was increasingly emboldened to fend for the manufacturers. The business communities also formed their own town-level associations for more specific local action. On the other hand, the Municipal Council was frustrated by the constant wars with the traders, which led to destruction of property and loss of revenue as well as giving the town a poor image, and by the government's apparent low prioritization of the town's development. Amidst all this, civil society networks in the city were working in parallel with the Local Authority, and without paying much attention to its activities. The central government departments also engaged in legitimacy wars with the Municipal Council, with proposals from the Municipal Council not being implemented and the central government, through the Ministry of Local Government, ensuring that it was in full control of urban development in Kisumu (UN-Habitat 2005a, 2005b).

Early initiatives in city-level knowledge co-production

Jomo Kenyatta Grounds (JKG) renewal process 2000–2002

An opportunity for change in urban development presented itself with the decision to celebrate the centenary of the town's founding. This was to be supported by the World Bank, which sponsored the construction of the Kisumu bus park, while Sida embarked on the renovation of the Jomo Kenyatta Grounds as a sign of its commitment to the Lake Victoria Region (Seman 2003). The Jomo Kenyatta Grounds were constructed as a model of a neutral, well-governed public facility offering environmentally sustainable services to the people of Kisumu. A process was carried out to identify and engage the key stakeholders in the town in the conceptualization, design, and integration of indigenous knowledge and its eventual translation into a facility that would reflect the input of the various stakeholders. This was the first deliberate attempt to involve the co-production of knowledge by different stakeholders in the development of a public facility. The main stakeholders (the council and government, civil society, the informal sector, academia, and the private sector) participated in the process of identifying the challenges and opportunities associated with the park. The design aimed both to preserve culture and to offer space for the public. A board of nine people representing the key stakeholders (MCK, CSOs, and private sector) was formed to run the facility (MCK 2007a, 2007b). This was a significant milestone in the centennial celebrations of the town and paved the way for the replication of this approach in the city.

The Kisumu City Development Strategy (CDS) process 2002–2004

Buoyed by the success of the JKG process and remembering the city's turbulent past, the Municipal Council of Kisumu engaged UN-Habitat to assist with city planning in 2002. This was then undertaken using the City Development Strategy (CDS) approach developed by the Cities Alliance, whose successes include Durban, South Africa, and cities in Eastern Europe (UN-Habitat 2005a). In partnership with the Intermediate Technology Development Group, UN-Habitat engaged the Municipal Council in the process under its Sustainable Cities Program (SCP), which aims to put people and environmental sustainability first in planning and development. A team was drawn from the JKG process to upscale their experience to the city level. Enumerators were engaged to conduct a city-wide census, engaging about 2,000 citizens. This formed the basis of the new planning for the city, which was conducted through a series of city consultations and working group sessions at the thematic and strategic levels. The process lasted two years, and culminated in the Kisumu CDS which was to provide the basis for investment in Kisumu for a number of years. The involvement of diverse stakeholders during the development of the CDS made it easy to use the strategy within the city, since the implementers and beneficiaries had contributed to the production of the plan, and were able to relate to the action areas within it. The success of the CDS is best illustrated by the fact that it served as the basic reference document for the Municipal Council in guiding development for several years.

Residents' associations 2003–2004

The continued clamor for citizen empowerment saw the birth of neighborhood associations in the leading cities in Kenya. In Kisumu, some individuals took advantage of this movement to amalgamate a few neighborhood groupings to form the Kisumu Town Residents Welfare Association (KITREWA). KITREWA aimed at mobilizing the residents to agitate for better governance in their neighborhoods within the city (MCK 2003). The residents of a given neighborhood would generate issues connected with improving their neighborhood. This noble initiative aimed at inviting all the stakeholders (from the community, government/local authority, academia, and civil society) in their private capacities as residents to participate in dialogue on ways of improving their living conditions. The initiative was successful in making the members aware of their rights and responsibilities, and resulted in more concerted neighborhood cleanliness initiatives. However, it was short lived, not going beyond the first phase funded by the World Bank. According to Merciline Oyier, the first program officer for the project, the initiative failed due to the lack of accountability occasioned by weak organizational structures with poor representation of the various stakeholders.

The neighborhood approach was adopted by Sana International, one of the local CSOs working with empowerment of the communities. Sana went on to help set up Neighborhood Associations in the unplanned settlements of Obunga, Manyatta-A, and Manyatta-B. These associations proved to be very useful in analyzing the challenges facing the three neighborhoods. A joint team formed by the Municipal Council of Kisumu and Sana led the process of training the community task teams in their thematic areas (Sana 2005, 2007). The initiative led to the formation of issue-specific task forces generated through dialogue. These included the Wandiege Community Water Service project, which serves a significant portion of Manyatta B informal settlement, and, in partnership with the Municipal Council of Kisumu, a mini demonstration Environmental Pedagogic Center located at Taifa Park (Obera and Ochieng 2004).

The Lake Victoria Region Local Authorities Cooperation (LVRLAC)

Lake Victoria Region Local Authorities Cooperation (LVRLAC) was formed in April 1997, "on the initiative of the Mayors of Entebbe, Uganda, Mwanza, Tanzania and Kisumu, Kenya, LVRLAC was established as a regional cooperation of Local Authorities in the LVR. The reason for the establishment was the awareness of the many varied environmental, economic and social challenges that are facing the Lake Victoria Basin and the recognition of the commonalities and similarities in the problems facing the Local Authorities around the Lake" (LVRLAC 2004: 2). LVRLAC borrowed from the model that had been developed in Kisumu in the reconstruction of JKG and the development of the CDS. It took advantage of its regional scope to involve players across the region, drawing on their experiences in co-production of knowledge to inform its strategies and implementation framework. The organization can therefore be seen as an example of a platform for co-production in the region. Two examples are given below of the results of this approach.

(A) CITY TWINNING

The twinning of Kisumu with other cities did not begin with the LVRLAC. Kisumu acquired sister city status with Roanoke, Virginia (USA), as early as 1976 and later with Boulder, Colorado (USA), as well as friendship status with Cheltenham (UK) in the 1990s, among others.

The formation and strengthening of LVRLAC as a network platform introduced a new dimension to city twinning. The member local authorities around Lake Victoria, spanning the three countries (Kenya, Uganda, and Tanzania), entered a new peer-learning and collective-action phase to sustainably improve the livelihoods around the Lake Victoria Region. Being a pioneer city, Kisumu took the leadership role in knowledge co-production within the region, having previous experience in the area from the JKG and CDS processes. The town became a learning center, which other players from the region would visit for benchmarking (UN-Habitat 2005a). The increased city twinning of local authorities around Lake Victoria earned them recognition and drew in cities from the North, especially around the Baltic Sea (LVRLAC 2010). In order to have more meaningful cooperation, Kisumu again took the lead in developing a framework for structured twinning with the North (MCK 2009), which was later adopted by LVRLAC. The town went on to cement other twinning arrangements with joint environmental projects such as the solid-waste management exchange with Nacka, Sweden, and the donation of a waste compacting machine from Porgrun, Norway (MCK 2012). In all these processes there was a constant deliberate effort to involve stakeholders whenever new knowledge was required to drive a process forward.

(B) ENVIRONMENTAL PEDAGOGIC CENTER (EPC) 2003–2009

Within the North–South and South–South twinning arrangement, one concrete co-production project, and probably the most interesting one, was the creation of the Environmental Pedagogic Center (EPC). The Union of the Baltic Cities (UBC) had a great deal of experience in sustainable development, having been employed to help rehabilitate and conserve the Baltic Sea (LVRLAC 2006). Based on the concerns raised about the status of Lake Victoria and the Region (Awange and Obera 2007; Carl Bro Consultants 2001), professionals from the UBC held a series of interactive meetings with their counterparts in LVRLAC and the local universities, which led to an agreement on the need for action to address the key challenges facing the Lake Victoria Region and the crafting of the EPC project modeled, after the

Swedish nature schools. Piloted in Kisumu and later spread to six other towns around Lake Victoria, the centers aimed to interact with the public on simple sustainable technologies, e.g. waste recycling, solar energy, solar water sterilization, gardening, and the use of human waste as fertilizer. The project had a scientific board comprising officials of the Municipal Council of Kisumu, Maseno University, and CSOs. This board would review simple, replicable, and innovative sustainable technologies to be displayed at the center (MCK 2005b). Members of the community would visit the center to learn while also making suggestions or bringing technologies that others could learn from (LVRLAC 2003).

The project was a success, leading one of the local NGOs to replicate it on a smaller scale in Manyatta unplanned settlement (Sana 2007). However, the operation of the center reverted to that of a normal council operation, as they could not sustain the interest past the funding phase. The board was therefore disbanded.

The Kisumu Action Team (KAT)

KAT I (2003–2004)

Having developed the CDS to guide the city's development, the stakeholders involved in the CDS process realized that the CDS and its participatory process were a departure from the normal top-down planning and development procedures used by the Municipal Council of Kisumu. Hence, there was a need for action to mobilize resources for implementation and monitoring, since the Council did not have the capacity to effectively carry this out on its own. It was therefore agreed that the CDS working group would be the first members of a team to steer the CDS. In a tri-sector meeting held on August 4, 2003 and convened by Ufadhili Trust, a local CSO based in Nairobi but run by people with roots in Kisumu, the increasing involvement of the three sectors (public, private, and civil society) in the governance of the city was recognized. However, the meeting noted several challenges with regard to the governance of the city:

1 lack of a shared vision among the city residents;
2 negative attitude among the citizens;
3 bad politics (opposition for its own sake);
4 uncoordinated stakeholder efforts; and
5 declining local economy.

The action team had a mandate to improve the image of the town, primarily focusing on visibility and impact, anchorage of the process for sustainability, and utilization of the CDS framework to exploit opportunities. At the fifth subsequent meeting, the action team noted the need to incorporate more members beyond the original six in order to add value to the activities of

Table 4.1 Membership of the first Kisumu Action Team 2003

1	Cllr. Clarkson O. Karan	Mayor/chairman
2	Mrs. Margaret Jobita	Town clerk and convener
3	Mr. C. S. Hayer	Rep. industry
4	Dr. Raphael Kapiyo	Rep. academic institutions
5	Mrs. Betty Okero	Rep. CSOs
6	Mr. Ben O. Kitoto	Rep. business community
7	Mrs. Anne Oketch	Rep. religious organizations
8	Mr. K. Binott	Rep. corporate sector–CBK
9	Mrs. Dolphine Oketch	Rep. women's interests
10	Ms. Merceline Oyier	Rep. youth
11	Mr. Simon Okeyo	Rep. informal traders
12	Mr. George Wasonga	MCK/Secretariat

the team. Members of the business community were then incorporated to the Kisumu Action Team (KAT). The membership of KAT I is shown in Table 4.1.

The team was launched on September 17, 2003, at which time the stakeholders offered equipment and their time to clean up the town. A series of meetings were held at which responsibilities were assigned to the various partners who would then provide feedback at the subsequent meeting. Issues addressed and goals achieved during the course of the year included:

1 Agreement on the theme and guiding motto "A Liveable City."
2 Identification and incorporation of key stakeholders.
3 Discussion on relocating the street traders from Oginga Odinga street to Ojino Okew and reorganizing the *boda boda* (bicycle taxis).
4 Community sensitization and engagement in the cleanliness of the town and premises outside town where the participants in the cleanup were based.
5 The Launch of the Kisumu Action Team with weekly activity highlights to constantly engage the locals and socialize them into taking voluntary action to better their living conditions.
6 Combining of projects into packages to seek support as well as secure partnerships to make them happen. This led to the packaging of all the parks and open spaces in the city into a single project to help seek partnerships for their upgrading. As a result of this, the Kisumu Lions Club supported the greening of the entire Jomo Kenyatta Highway as well as the development of Jamhuri park. ICRAF redecorated a roundabout at the Kakamega road junction opposite the Kisumu Boys School, and the World Vision installed litterbins within the town, to name a few examples.

The realization by stakeholders that the Lake Victoria Basin Commission (LVBC) headquarters was to be in Kisumu served as an impetus for sustained action to improve the city. The various activities earlier identified in the CDS were implemented and the city slowly began to take a new shape. Two issues that provided an arena for demonstrating the building of a platform for co-production of knowledge concerned the *street traders* and the *boda-boda* cyclists.

From the perspective of KAT, street trading was a growing concern due to its disorganization and the conflicts it involved, and hence it needed to be reorganized. This was all the more so for Oginga Odinga street, where the hawkers took over the walkways and entrances to formal businesses. KAT recommended that they be relocated to Ojino Okew, a backstreet. Consultative meetings were held with the street traders, during which teams were formed to oversee the transition (MCK 2003). A team comprising representatives from the social services and MCK's environment departments carried out a census of the traders. The engineering and planning departments were to plan and mark out Ojino Okew street, which had been designated a hawkers street under the previous JKG process. However, this was unsuccessful due to numerous challenges that arose during the implementation. Nevertheless KAT led to an increase in citizen involvement in the planning of the city. First, the street traders objected that the designated street could not accommodate all of them. Second, they demanded that the new street should first be developed, with sheds being put up, before they could consider a move. They later accused their secretary, who represented them on the KAT, of conspiring against them, and thereafter were unwilling to cooperate in the process. On the other hand, the team set up by the Council to oversee the process also took sides against the traders. This relocation exercise turned chaotic, with the councillor representing the affected traders leading a public demonstration against the exercise, among a host of other issues. While these challenges held back the implementation of the plan, the situation on the streets deteriorated further as more traders moved to both Ojino Okew street and Oginga Odinga street. These moves were motivated by their desire to be in a strategic position from which they could be the first beneficiaries in any relocation process (MCK 2005b). Between the political undertones in the process, MCK's internal turf wars, and perhaps a need to leave the numbers unknown so that the revenue collected remained unknown, the relocation has not been carried out to date. It is also telling that the street traders resisted an attempt to group them with the team trading in Oile park, saying that the Oile park team was an extension of some MCK officials' businesses or their beneficiaries, and would therefore have an unfair advantage over them.

Representatives of the *boda-boda* were never part of the Kisumu Action Team, although their issues were often discussed. An opportunity presented itself during the Kisumu Sustainable Urban Mobility (SUM) project facilitated by UN-Habitat under the auspices of the CDS. The project was to address some of the concerns about transport organization discussed by the Kisumu Action Team and in the studies carried out by a UN consultant in which the processes used in JKG were replicated in the strategy development (MCK 2004c; Opiyo 2005). The consultant's work provided a very good foundation to guide the stakeholders in the process of evidence-based planning, as it presented data on traffic volumes by modes, means, time of day, and routes. This kicked off a series of consultations within the no-cost and low-cost components of the project, incorporating revisions from consultations which were to address the issue of regularization of the *boda-boda* as well as relocation of the traders at Oile market (MCK 2003, 2004a, 2004b, 2005a, 2006a). The process of implementing the project dragged on, however, for a host of reasons. For instance, the reorganization of the bicycle taxis required that drivers be registered, licensed, and uniformed, and display clear routes—to prevent criminality and provide revenue to invest in further infrastructure development. However, the *boda-boda* members argued that licensing them for a fee would take them back to the old days when they would be arbitrarily arrested for not purchasing a sticker from City Hall. However in 2006, a compromise was reached, with concessions leading to an agreement on the content of the by-laws. The delays in the enactment of the by-laws coupled with the lack of accountability in the implementation of the physical infrastructure led to yet another change in leadership of the *boda-boda* grouping, the previous officials having been perceived as compromised. However, according to minutes of the SUM working group meetings, these by-laws were to be published in the *Kenya Gazette* with amendments to include the motorbike taxies, which were steadily eating into the bicycle taxis' market (GoK 2008). Though delayed, the process resulted in very useful consultations between the *boda-bodas* and the Municipal Council of Kisumu. The officials of the *boda-boda* were included in the subsequent KAT meetings.

KAT II (2005–2007)

In the period between the years 2005 and 2007, the KAT was loosely referred to by various names—KAT, KAT Steering Group, and Kisumu Action Team Development Forum—in the understanding that the small select team was acting on behalf of the larger stakeholder groups. However, the initials KAT stood out.

KAT's development was slowed down, however, by the 2004 change of city leadership, from Mayor Clarkson Karan to Mayor Priscah Misachi, with the accompanying high turnover of the chief officers, especially the town clerk (chief executive officer), of the Municipal Council. This was compounded by the long time required to sell the KAT idea to the new mayor as a vehicle to help her achieve her goals for the city, as the idea required packaging and she needed to be socialized into the new paradigm. For this reason, KAT went through a period of low activity (2004 until late 2005). The Working Group on the City Development Strategy continued holding public consultations with the support of UN-Habitat and Intermediate Technology Development Group (now Practical Action) leading to an agreement on the thematic areas for challenges facing the city, a city vision, and the goals leading toward realizing this vision. Their recommendations were discussed and the full council of the MCK finally adopted the CDS.

The CDS vision was "A leading transportation, communication and commercial hub in the Great Lakes Region offering great tourism and agro-investment opportunities" with the goals being to:

- Improve and conserve the urban environment for the sustenance of the lake and its resources
- Empower citizens and promote active stakeholder participation and involvement at all levels of local governance
- Improve access and provision of basic infrastructure and health services
- Promote sustainable socio-economic development in line with the national economic recovery and poverty reduction strategies
- Institute a planning culture that is responsive to the needs of the urban populace, while actively engaging their participation in providing strategic direction

(UN-Habitat 2005a: 22)

As a result of the Council's adoption of the CDS, a new KAT steering group was formed. Its membership was revised, leaving out the informal sector (possibly due to the distaste caused by their non-cooperation) and including more boardroom-savvy private sector and senior government representation. This may have been a reflection of the mayor's impatience with the troublesome informal sector. At their inaugural meeting April 28, 2005, KAT agreed on the need to work toward implementing the CDS by, among other things, setting up task forces to deal with the issues identified while seeking other members to support the cause. The new KAT was formally launched on May 25, 2005, with members as shown in Table 4.2 (KAT 2005).

At the subsequent meeting, it was reaffirmed that KAT would provide a development platform that would support the growing urban development challenges as well as create the right synergies for consolidating gains and promoting coexistence among the different development actors (KAT 2006). The meeting took cognizance of the increasing political goodwill for Kisumu's development and the realization that the growth of poverty was a time bomb. At the same time, it was realized in KAT meetings that some members were not yet conversant with the CDS as the guide for development of Kisumu town. On the other hand, the commitment of the membership came into question, as some members habitually failed to attend meetings. Issues of governance were constantly brought up at KAT meetings, especially the non-responsiveness of the Council to issues raised by KAT, touching mostly on planning and environment. Subsequent meetings were to call for much more council involvement in the KAT committees so as to get speedy action. Action plans, for implementation by the Council, were developed by KAT and included issues of planning and creating land banks for development.

Table 4.2 Membership of the second Kisumu Action Team 2005–2007

1	Cllr. Prisca M. Auma	H/W the mayor
2	Mr. Wisdom K. Mwamburi	Town clerk
3	MPs	Kisumu Town East & West
4	District Commissioner	Kisumu District
5	Mr. C. S. Hayer	Rep. industrialists
6	Prof. F. Onyango	Vice-Chancellor Maseno University
7	Chairman, Law Society of Kenya	Western Region
8	Mr. Sunil Shah	Kenya Ass. of Manufacturers
9	Mrs. Grace Onyango	Senior citizen (First African Mayor of Kisumu)
10	Mr. D. P. Mandavia	Rep. business community
11	Mr. George Olilo	Chamber of commerce
12	Manager, CBK	Rep. financial sector
13	Mayor's chaplain	Rep. religious organizations

It is worth noting that with KAT having been dubbed an action team, the Kisumu Mayor was possibly looking for a membership comprising stakeholders at a decision-making level, which could explain the exclusion of a number of stakeholders. As such, the knowledge co-production was limited to the ability to mobilize resources for action. The year did however see the first formal engagement of academia and practitioners in the generation of knowledge on the slum situation in Kisumu (UN-Habitat 2005b). A team led by George M. Onyango from Maseno University and planning professionals from Maseno University, Sana, and the Municipal Council of Kisumu's Environment and Planning departments conducted the Situation Analysis of the Informal Settlement. The composition of the team was useful in that the team member's experiences in community engagement enabled them to identify challenges facing the community in an objective manner. It is worth noting that local authorities had been shown to have challenges in the consultation processes and implementation of projects under the Local Authorities Service Delivery Action Plan (LASDAP) process (Owala, undated). The study was to form the basis for slum upgrading and poverty alleviation under the Cities without Slums (CWS) program of UN-Habitat (UN-Habitat 2005b). However, there is usually a disconnect between research and implementation. The inclusion of the practitioners in the exercise therefore helped the researchers appreciate the process as a means to achieve better planning as well as to be able to replicate it in similar situations.

KAT was not so active during this period, however, as complaints had been voiced about the Council's lack of commitment to the initiative. These concerned the apparent lack of appreciation on the part of the political leadership and the continued high turnover of the chief officers. What remained of the team (mainly council department representatives) focused mainly on some of the issues raised earlier while looking at new opportunities, though with no meetings. Nevertheless, the foundation had been laid for action to address many of the issues raised. The challenge was to gather the resources for further action and the capacity to build on this foundation to create a formidable team for translating co-produced knowledge into action.

Millennium city status

The increasing poverty levels, deteriorating environmental conditions, and associated challenges that long had plagued the developing countries became a global concern in the new millennium. This led to the 2000 World Summit that set the Millennium Development Goals (MDGs) with a promise for action by the participating heads of state. Each state was to develop an appropriate framework to fight poverty, hunger, illiteracy, poor health, and environmental degradation. Prof. Jeffrey Sachs, the then special adviser to the UN Secretary General on the MDGs, established an intervention to specifically address the MDGs, the Millennium Villages Project (MVP), under the auspices of the Earth Institute of Columbia University (MCI 2007a, 2007b; MCK 2006b, 2007a).

In 2006, the African Millennium Cities Initiative was formulated to assist medium-sized cities in achieving the Millennium Development goals. The experience from the Millennium Villages Project came in handy in implementing the MCI as a learning platform. One of the most successful MVPs was Sauri, some 60 kilometers from Kisumu. Kisumu was thus seen as the best-suited partner to the Sauri village. Kisumu had already made efforts to address the MDGs, as it had already established a basis for dialogue by stakeholders on the MDGs within the framework of the City Development Strategy. It is on this basis that on January 11, 2006 Prof. Sachs declared Kisumu the first Millennium City (KAT 2006). A staffer from the Earth Institute was seconded to the town to begin the process of engaging the stakeholders toward the setting up of the initiative.

KAT responded to this declaration by crafting a framework under which the initiative would fall, thereby making it easier to achieve the MDGs. The structure aimed at inculcating stakeholder participation to enhance ownership, and hence action toward attainment of the goals. The Environment Committee and the Municipal Council of Kisumu later approved this in 2006.

The year 2007 was full of expectations, with the various stakeholders keen on seeing what the international community had to offer with regard to the town's new status. A link was established with the Ministry of Planning, under which the National MDG secretariat was hosted, as well as with the UNDP, in which the Finnish government's support for the MDG programs was anchored. A series of studies were commissioned, the mayor appeared at publicity events, and a major investment conference was held in Kisumu. The Millennium Cities Initiative adopted a two-pronged approach: one focusing on promoting investment, which led to the establishment of the Kenya Investment Authority office in Kisumu; and the other concentrating on the social sector, which carried out MDG needs assessments so as to help plan for their achievement as well as to seek partnerships to foster further

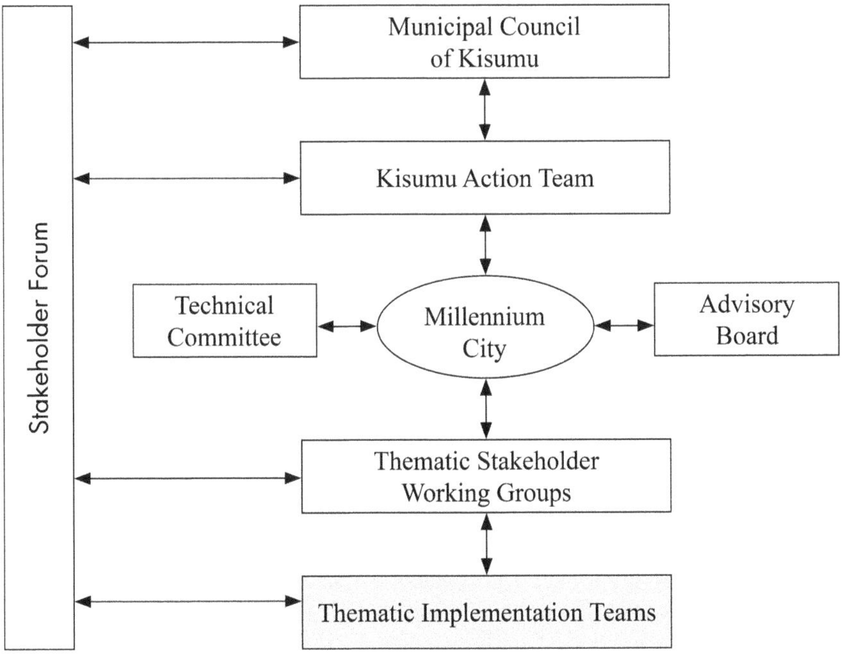

Figure 4.1 The Kisumu Millennium City Framework
Source: MCK (2007).

development. The Millennium Cities Initiative focused on supporting under-resourced cities by conducting research and policy analysis with the help of a huge number of research students, professionals, and partners (see Figure 4.1). This gave birth to a series of MDG needs assessments on health, education, gender, and water and sanitation. The organization also facilitated a research project on investing in Kisumu (MCI 2007b). From this research and policy analysis work, the organization was to move onto the next stage: to involve the city stakeholders in reviewing the City Development Strategy and identifying the resources needed for its implementation (MCI 2012).

The work of MCI builds on the Kisumu City Development Strategy, which came to an end in 2009. While the CDS relied on a "city scan," a rapid but comprehensive city-wide survey, which identified the issues affecting the town, it lacked concrete figures and at best relied on some amorphous percentages, making it difficult to monitor the achievement of the set targets. The 2004–2009 CDS also lacked strategies for economic development. The new effort spearheaded by MCI was built on the framework established through the CDS. It engaged academia to carry out needs assessments and investment profiling for specific sectors of the city's economy, drawing on the practices that had led to the successful development of the CDS. The initial studies covered the social sectors, namely education, health, and housing. The objective was to contribute to the drafting of a new development strategy for Kisumu.

KAT III (2008–2012)

Being an election year, 2007 saw an impending change in the political leadership throughout the country and put the brakes on the work of KAT. As usual the political leadership focused more on their own survival and avoided tough negotiations and actions for the sake of the electorate which might jeopardize their chances of retaining their positions. This was to be compounded further by the fiasco of the 2007 election and the accompanying violence that lasted over two months. The city experienced destruction of property and loss of life on a scale not seen before. National peace negotiations finally allowed for a return to some level of normalcy in Kisumu. Dr. Kofi Annan, former UN Secretary-General, led a team of eminent persons to negotiate a truce between supporters of Raila Odinga and Mwai Kibaki, resulting in a coalition government with Kibaki as President and Raila as Prime Minister. The Kisumu Municipal Council was convened with Sam Okelo being elected mayor. The new mayor was briefed on the CDS and the processes that had been going on under KAT (MCK 2008a, 2008b, 2008c). The mayor convened the first stakeholder meeting on March 19, 2008 (KAT 2008). The meeting was well attended, with over 123 registered participants from all stakeholder groups in the town. At the meeting ten major issues were identified as needing urgent action:

1 Accountability on the part of the council.
2 Reliability in service provision by the council (waste, water, roads, housing, among others).
3 Agreement that the Council lacked capacity and needed support.
4 Improving planning and implementation.
5 Improving the business environment.
6 Creating employment for the youth.
7 Creating space for everyone to live and carry out their socio-economic activities.
8 Urgently addressing the severe conflicts of interest among council officials.
9 A more responsive council.
10 Regular forums for sharing/feedback and evaluation.

A team comprised of eight individuals was selected to look at the issues raised with a view to developing short-, medium-, and long-term plans. These were to be drafted as a concept for the reconstruction of the town after the massive losses due to the post-election violence. The Millennium Cities Initiative took the role of the secretariat to guide the process.

The team did a quick review of the stakeholders' meeting and drafted a framework, as well as compiled a list of representatives to be included in the renewed Kisumu Action Team. The new participants included:

- the mayor;
- two Municipal Council of Kisumu representatives;

- three representatives from the Informal sector;
- three representatives from the CSOs;
- five representatives from the private sector;
- a representative from the faith-based organizations;
- four representatives from government departments; and
- three representatives from academia.

The new membership was to include twenty-two members, though only about fifteen were active. Letters of appointment were prepared for the selected members, who were then invited to the first KAT meeting on April 2, 2008. At the meeting it was agreed that the terms of reference were to provide support to the leadership of the city as well as create a framework for public–private partnership, with measures to:

1. Identify strategic action areas.
2. Link the strategic areas with ongoing initiatives.
3. Develop strategies for execution.
4. Mobilize resources.
5. Support the Municipal Council of Kisumu in its work of implementation.

The meeting identified the short- and medium-term action areas. At the fifth meeting on April 24, 2008, the team agreed to set up three working groups to address the immediate issues (traffic reorganization, resource mobilization, and environment). The KAT working groups subsequently formed four subcommittees, as approved at the sixth KAT meeting on May 7, 2008. These were:

1. Health and environment
2. Planning subcommittee
3. Local economic development
4. Infrastructure subcommittee

These were to address the main issues emerging from the stakeholder consultations.

The subcommittees went on to engage various institutions and stakeholders in these endeavors. One of the greatest achievements of the KAT at this stage was the development of the Kisumu Reconstruction Strategy which the mayor took up and marketed to the various partners, resulting in the French government's decision to fund the Kisumu Urban Project (KUP) to an amount of Ksh 4 billion (4 million euros). The news of these achievements was shared with the general public at the stakeholder forum held on April 28. Kisumu Municipal Council also had a poor history of project implementation. For instance the traffic reorganization plan was developed and approved on April 1, 2008. Its implementation was not effected until 2012, by which time some of the members had stopped coming to the meetings. In 2009, for example, only one meeting was held, on March 2, 2009, at which the Council briefed participants on the progress on the KUP project as well as other agencies that had supported the city. These presentations continued at the subsequent meetings in 2010, which some members viewed as window dressing for the donors who attended these meetings. This trend was to go on into 2011 and 2012, only this time with a new project under the leadership of a new group from the KAT taking the lead and steering action in this direction.

Lessons learned from the KAT process

It is evident that the KAT processes outlined above bear witness to the changing demands and conditions that influence the success of different governance constellations. All facets of governance have one way or another been an issue in the KAT processes. These include the following.

Timeliness: there were constant complaints about timeliness, including the lack of a clear schedule of meetings. Even in cases where the members agreed on the date and length of the meeting, these were not followed. Meetings would be convened at the convenience of the leadership and sometimes took longer than planned and had changing agendas. Increasingly, members of the KAT would send their apologies instead of attending.

Equity and representation: while there was a desire to incorporate all the social groupings of the city in the running of its affairs, truly inclusive representation in the team was a challenge to achieve because of the small size of the KAT membership. This created a dilemma concerning which stakeholders to choose to include in the KAT membership. For instance, the informal sector is the largest sector in the town, however, it was not possible to have a commensurate number of its representatives in the KAT, as the operations of the team would have become untenable. On the other hand, having too few representatives from the informal sector caused problems, especially at the action stage, with friction and splitting of the sector. The street traders split into many groups, and at some points it was not clear who should be invited to the meetings. On the other hand, some segments of the KAT membership felt that their contributions were not given much consideration, because their suggestions often were overlooked in reporting to the leadership, and because of the long delays between proposals and action. For instance, although the traffic reorganization plan was developed in 2008, it was not until 2012 that an attempt was made to implement it.

Accountability: horizontal and vertical accountability challenges have been a key issue and perhaps one of the most visible and emotionally charged aspects dealt with during the near decade of KAT's existence. Horizontally, members were at times unable to account for their time spent in the meetings and whether successful outcomes could be attributed to their contributions. Sometimes there was a feeling that decisions were made without their participation. Vertically, the forum for stakeholder feedback, while factored into the operation, was not convened on a regular basis or responsive to issues that were previously raised. This lack of holistic feedback caused problems for the stakeholder representatives at KAT. At times there was a feeling that KAT was a "cooling house" between the city authority and the stakeholders on the ground. Whenever a hot issue emerged, it would be passed through KAT to demonstrate that the Council was taking action, hence the feeling by some members that they were working at the behest of the MCK.

Communication: like all stakeholder issues, communication problems dogged the KAT operation over time. The timeliness and modes of communication were constant issues, with meetings sometimes called overnight via text messages, which did not go down well with most members. Members also raised concerns about disparities between communications with the mayor and with the membership, and felt that results of deliberations meant to reflect the interests of stakeholders would be moderated or omitted from the communication with town hall.

The county government and the demise of KAT

The new Constitution introduced county governments, with Kisumu City falling under the County Government of Kisumu. The position of mayor was abolished and the management of the city was transferred to a City Management Board with a city manager handling the day-to-day operations.

As a result of the changes, new power centers have emerged in the governor's office and the County Assembly. The prime function of KAT has lost its relevance in a system where the city has become just one of several players in a region competing for resources. Decision-making has shifted to a whole new team who do not seem to see a role for KAT in the new dispensation. With no anchor in the city establishment, KAT became history, and the players have repositioned themselves in new roles.

Kisumu Local Interaction Platform (KLIP)

The Kisumu Local Interaction Platform (KLIP) was established through the work with Mistra Urban Futures. The platform has grown in stature and strategy as it positions itself to take over the space left by KAT. There are two projects areas that are active at present: Market Places and Ecotourism. The activities connected to these projects include collaboration between the researchers from Maseno University, Jaramogi Oginga Odinga University of Science and Technology (JOOUST), the University of Gothenburg, Chalmers, CSOs, the County Government of Kisumu and surrounding counties, community members, and a number of local institutions. A number of students are also registered for their doctoral studies with support from the project and are at various stages of study.

The implementation of both the Market Places and Ecotourism Projects are taking place in a very particular, rather favorable, context in Kenya. Institutional reforms are ongoing under the framework of the new Constitution. The urban development sector has benefited from an increased focus among the national priorities. The Kisumu Urban Project, pilot project for broader urban development programs, has been launched. Kisumu has finally officially been granted the status of City and its leaders are extremely keen to make the most of both these initiatives to fulfill their vision. Last but not least, the projects are taking place during the first phase of devolved government with the governor keen on spreading benefits beyond the city confines.

The Market Places team, consisting of a number of master's and Ph.D. students from both Maseno and Chalmers, work on a variety of topics including business vulnerability, solid waste management, the role of ICT in periodic markets, and the accessibility of market stalls. One example of a specific project is the work being done on solid waste management in informal settlements. While working with Civil Society in Manyatta estate on waste separation, knowledge has been co-produced between students and residents on different types of waste management needs. This initiative has enabled the resident association to engage in solid waste management and improved both the environment and increased business opportunities. This system of waste management is designed to be replicated in other markets within Kisumu County and even beyond. Waste management is also one of the flagship projects that will be funded under the Kisumu Urban Project being implemented by the County Government of Kisumu.

The KLIP ecotourism Ph.D. student core group with members from the University of Gothenburg and JOOUST have conducted collaborative work in Dunga Beach, on the shores of Lake Victoria, involving community organizations and residents in activities such as workshops about identity, packaging, and prototyping. The aim of the project is to develop the ecotourism site in a sustainable manner, focusing not only on economic gains but also on environmental and sociocultural dimensions. The students are explicitly working with transdisciplinary approaches to knowledge co-production and integrating them with their respective fields of tourism, marketing, resource management and design. For example, two test tours were held where Ph.D. students, tour guides, and residents jointly prototyped two guided tours as well as a waste collection point and signage system. The residents are fully engaged in the processes of knowledge production as part of the tour as storytellers, as craftspeople engaging the tourists in rope- and papermaking, and food preparation with the tourists.

These examples of researcher-initiated co-production have met with a number of challenges within the KLIP environment. There are challenges working within this multidisciplinary and cultural environment that requires consistent interaction and sharing of knowledge and skills to enhance learning. The joint researchers between Kenyan and Swedish students also have the challenge of distance. Use of Skype and other cloud computing activities have to some extent helped the students interact. There is a clear need for participatory research approaches that benefit practitioners and increase local knowledge, though this is a tedious and time-consuming process. However, involvement of the community is allowing for co-production of knowledge and giving insights into community dynamics at various levels. Organizations and institutions are willing to learn and have therefore created links that will enhance the activities of the KLIP team within research, policy, and implementation.

Conclusions

As noted in Chapter 1, co-production of knowledge for sustainable urban development is a nonlinear, collaborative process. In Kisumu the process started with peaceful coexistence and minimal conflict between various players in the city. The need for working together led to more actors coming into play, generating some mild conflicts. As the stakes grew in this process of city development there emerged major conflicts that saw shifts in strategies and increased engagement of various stakeholders in co-production for city development. Interdisciplinary, cross-sector, and policy approaches have been used to solve some of the urban planning and development issues in the city, drawing on input from various stakeholders in the city. This provided a democratic basis for decision-making and made negotiations possible between different interests. This is what has been realized over the years, as best exemplified by the KAT, and it is becoming more objective and genuine with time. For example, the first KAT comprised the city elites and excluded the informal sector, and had the primary objective of raising resources to address the challenges facing the city. The second KAT adjusted this slightly to include heads of academic institutions and government departments. The third KAT, however, adopted a more objective and realistic approach to the co-production of knowledge by including a greater diversity of members. This is also reflected in the products of the three KATs. KAT I was able to mobilize local resources around the greening of the parks and road islands, as well as get trucks from the private sector to haul away garbage heaps. It thereby stirred local action and attracted international recognition. It began however from the premise that the informal sector was the cause of the problem and not part of the solution to the city economy. This explains why it took such a long time to work with the *boda boda* and the street traders.

KAT II shifted the work a notch higher, to oversee the engagement of the informal sector, though not as equal partners in the city decision-making process. The informal sector was left out, possibly due to its "nuisance" tag. KAT II stepped up the outside recognition to a higher level leading to the adoption of the Kisumu City Development Strategy and the declaration of Kisumu as a Millennium City in 2006. The stakeholders' contributions to the beauty and cleanliness of the city led to further projects like litterbins, street lighting, and regular clearing of garbage heaps. However, being composed of top officials from different institutions made it difficult for KAT II to analyze issues in-depth and possibly to oversee the implementation of agreed plans, which had to be left to the Municipal Council of Kisumu. For this reason, the

City Development Strategy remained largely unimplemented. The second KAT was therefore weak in co-production of knowledge, and accordingly also weak in evidence-based implementation and monitoring.

KAT III had the advantage of hindsight, and hence was able to address some of the challenges of engaging stakeholders in co-production in the dynamic city planning environment. The team included high-level professionals as well as representatives of the informal sector, and divided them into sub-committees in accordance with their technical capacities. For this reason, the third KAT was probably the best of the three in terms of the co-production of knowledge. For instance, at meetings the representatives of the street traders presented valid arguments for their being on the streets. The academics and other professionals listened to the traders, gave the problem serious thought, and suggested constructing aesthetically acceptable kiosks for their street trade, which would also improve security. The same applied to the transport reorganization. These plans were agreed on and shared with the implementing agency (MCK), though it took time to win acceptance for their implementation. Furthermore, reviewing the post-election challenges, KAT III came up with the reconstruction strategy, which ultimately was to form the basis for the largest infrastructure project ever realized for Kisumu, the Kisumu Urban Project, funded by the French government.

As this history of Kisumu shows, different types of knowledge co-production are part of a continuous learning process. The composition of the Kisumu Action Team changed over time, with some stakeholders who felt the team did not meet their best interest dropping out, and it became clear that certain governance issues had to be addressed to enable the team to proceed. This led to the realization that all stakeholders were important, which resulted in the final composition of the membership, one that represented the different sectors of the city fairly.

In the last couple of decades, different types of knowledge and expertise from a variety of actors have influenced urban planning and policy-making in Kisumu. This has mainly been due to the way researchers, practitioners, and citizens have worked together to co-produce knowledge to contribute to solving the complex challenges of developing Kisumu. Issues of sustainable urban futures have been addressed as the various players working as teams have formulated strategies to propel the city forward and guide it toward the ideal of sustainable growth and development. We see examples in which planning processes have been deliberately guided by researchers with the involvement of practitioners as they work with citizens to evaluate alternative urban futures. KAT was a perfect vehicle for this process, one of its strengths being the participation of various stakeholders with a clear mandate. Academics and practitioners, as well as policymakers, have contributed to knowledge production, with UN-Habitat acting as the macro-policy developer driving a number of the processes geared to solving the complex societal problems linked to Kisumu's urban development. UN-Habitat has also invited a number of researchers, practitioners, and policy-makers to scrutinize the interventions proposed for Kisumu from the point of view of various ideologies and beliefs to create opportunities for debate and action. The involvement of practitioners, researchers, and policy-makers working in a transdisciplinary environment has created an opportunity for Kisumu to explore the heterogeneity of scientific practice (Hessels and van Lente 2008).

Participation and capacity-building have been the cornerstones of knowledge co-production in Kisumu. No single actor has the capacity or power to grasp or address the full complexity of urban development. KAT demonstrated this fact best, and created opportunities for dialogues that enabled the team to bridge gaps between rich and poor, local and global, and present and future generations. The activities at the Kisumu Local Interaction Platform (KLIP) of Mistra Urban Futures have built upon this framework. Driven by universities, KLIP has positioned itself as a key driver in co-production of knowledge. The future of the university is being shaped by the transition to a more knowledge-intensive society, an environment where universities will be able to occupy a more central and powerful position. Adopting an evolutionary model of the university, one can argue that different "species" of universities have emerged and coevolved over the centuries. These species place different emphasis on the functions of teaching, research, and contributing to the economy, and have different ways of relating to the changing relationships and tensions between these functions (Martin and Etzkowitz 2000). Maseno University has gradually positioned itself within Kisumu as such a university, and in working with UN-Habitat and later on through various partners in KLIP such as JOOUST, Manyatta Wastepickers and other CSOs, it has found its niche as a facilitator of knowledge co-production and a consumer of the knowledge so generated.

Despite the considerations stated above, the authors are of the opinion that co-production in urban development and management is only valuable if the resulting knowledge is translated into policy for development—with the various stakeholders taking part in the furtherance of the ideals—and into better physical, economic, and psycho-social living conditions. In Kisumu there is evidence of the efficacy of stakeholder participation that to a great extent has influenced the development of strategy and policies in the last two decades. The work of UN-Habitat, Maseno University, Municipal Council of Kisumu, KLIP, and CSOs has led to the adoption of a transdisciplinary approach to knowledge co-production, as evidenced by a number of planning and policy documents.

In view of the new Constitution's transfer of many governance areas and functions to the county level, opportunities abound for new and enhanced co-production of knowledge for county development given the fact that most of the functions are new, with no precedents on operationalizing them. However, it is important to realize that Kisumu's experiences with the Kisumu Action Team framework for stakeholder participation form a major platform for addressing issues of sustainable urban development and citizen and institutional participation, and for ensuring that policy development is built on co-produced knowledge.

References

Anyumba, G. (1995) *Kisumu Town: History of the Built Form, Planning and Environment: 1890–1990*. Delft, Netherlands: Delft University Press.

Awange, J. L. and Obera, B. O. (2007) 'Motor vehicles: Are they emerging threats to Lake Victoria and its environs', *Water, Air and Soil Pollution*, 182(1): 43–56.

Carl Bro Consultants (2001) *Study of the Environmental Situation in Three Urban Centres of Kisumu, Kampala and Musoma*. Stockholm: SIDA.

Government of Kenya (GoK) (2008) The Local Government Act Chapter 265.

Hessels, L. K. and van Lente, H. (2008) *Re-thinking New Knowledge Production: A Literature Review and a Research Agenda*. Innovation Studies Utrecht (ISU) Working Paper Series. Copernicus Institute for Sustainable Development and Innovation, Utrecht University.

Kisumu Action Team (KAT) (2005) *Minutes of the KAT Meeting Held on May 25, 2005*. Kisumu: Kisumu Action Team.

——(2006) *Minutes of the KAT Meeting Held on January 20, 2006*. Kisumu: Kisumu Action Team.

——(2008) *Minutes from the Stakeholder Meeting Held on March 19, 2008 at Aga Khan Hall*. Kisumu: Kisumu Action Team.

Lake Victoria Region Local Authorities Cooperation (LVRLAC) (2003) *Environment Pedagogic Centre (EPC) project*. Entebbe, Uganda: LVRLAC.

——(2004) *LVRLAC Strategic Plan, 2004–09*. Entebbe, Uganda: LVRLAC.

——(2006) *Annual Report*. Entebbe, Uganda: LVRLAC.

——(2010) *Strategic Plan 2009–2014*. Entebbe, Uganda: LVRLAC.

Lonsdale, J. (1989) 'The conquest of state 1895–1904', in W. R. Ochieng (ed.), *A Modern History of Kenya 1895–1980*. London: Evans Brothers.

Martin, B. R. and Etzkowitz, H. (2000) 'The origin and evolution of the university species'. Paper presented at the workshop Organisation of Mode 2/Triple Helix Knowledge Production, held at Goteborg University, October 20.

Millennium Cities Initiative (MCI) (2007a) *Attracting Investment to Kisumu: Opportunities and Challenges*. New York: Earth Institute, Columbia University.

——(2007b) *Investing in Kisumu*. New York: Earth Institute, Columbia University.

——(2012) *Spurring Sustainable Urban Development: Effecting Lasting Change in the Millennium Cities*. New York: Millennium Cities Initiative.

Miller, C. (1971) *The Lunatic Express: An Entertainment in Imperialism*. London: Macmillan.

Mitullah, W. V. (2004) *A Review of Street Trade in Africa Draft*. Commissioned by Women in Informal Employment Globalising and Organising [WIEGO]. Cambridge, MA: Kennedy School of Government.

Municipal Council of Kisumu (MCK) (2003) *Proceedings of the Consultative Meeting with Street Traders Held at Social Hall*. June 10.

——(2004a) *The Kisumu City Environmental Profile on Sustainable Urban Mobility.*

——(2004b) *Project Document for Sustainable Urban Mobility (SUM) Demo project.*

——(2005a) *Minutes of the Environment Committee Meeting Held on the 14th March 2005.*

——(2005b) *Minutes of the Planning Committee Meeting Held on the 28th July 2005.*

——(2006a) *Minutes of the SUM Working Group of 27th November 2006*

——(2006b) *Institutional Framework for Millennium City.*

——(2007a) *Kisumu Millennium City Institutional Framework.*

——(2007b) *Minutes of the Special Full Council Meeting Held on the 14th March 2007.*

——(2008a) *Minutes of the Full Council Meeting Held on the 14th March 2008.*

——(2008b) Non Motorised and Motorised Cycles By Laws 2008.

——(2008c) *Minutes of the Full Council Meeting Held on the 14th March 2008.*

——(2009) *Establishing Sister City Affiliation Policy for the City of Kisumu.*

——(2012) *Environment Committee Report.*

Obera, B. and Ochieng, J. (2004) 'A five day training report of Manyatta A and B on Integrated Solid Waste Management for Sustainable Development among the Urban Poor Communities', Kisumu.

Obudho, R. A. and Waller, P. (1976) *Periodic Markets, Urbanization and Regional Planning: A Case Study of Western Kenya.* London: Greenwood Press.

Onyango, G. M. (1998) 'Regional Organizations: Conflict in urban planning in western Kenya', in *African Development Perspectives Yearbook 1997/1998.* Bremen, Germany: University of Bremen.

Onyango, G. M. and Ogonda, R. T. (1999) 'The port and town of Kisumu: Past, present and future', *African Urban Quarterly* (special issue on port cities), 9(1&2): xx.

Opiyo, T. (2005) *Urban Mobility Conditions, Problems, and Recommended Policies in Kisumu.* Nairobi: UN-Habitat/ IHE-UNESCO/ITDG.

Owala, C. (undated) Management of Devolved Funds: A Case Study of Kisumu Municipality. A network of civil society organization, under the leadership of Community Initiative Action Group-Kenya with Facilitation from Ufadhili Trust.

Proceedings of the Jomo Kenyatta Grounds Rehabilitation process between 1999 and 2002.

Sana (2005) *Sana International Annual Report for 2005.*

——(2007) *Sana International Annual Report for 2007.*

Sandbrook, R. (1975) *Proletarians and African Capitalism: The Kenyan Case, 1960–72*, Cambridge: Cambridge University Press.

Seman, P. O. (2003) *Project Document for an Integrated Waste Management Plan for Kisumu Town.* Kenya.

Ufadhili Trust (1997) *The Kisumu We Want, A Report on the Process.*

UN-Habitat (2005a) *Cities Development Strategies for Improved Urban Environment and Poverty: Kisumu City Development Strategy.* Nairobi: UN-Habitat.

——(2005b) *Situational Analysis of the Informal Settlements in Kisumu.* Nairobi: UN-Habitat.

UN-Habitat/UNEP (1999) *Preparing the SCP Environmental Profile.* An SCP Source book series. Nairobi: UN-Habitat Nairobi.

5 Lessons on the research–practice relationship

From critique to co-production in Greater Manchester

Beth Perry and Tim May

Introduction

In the broad field of urban studies, insufficient consideration has been given to the relationship between the city and how urban knowledge is produced. Often urban scholars will describe their research processes, but without investigating how their methods have shaped the knowledge that they produce. Reflexivity about the links between research, policy, and practice has been *one-way*, primarily concerning the relationship between individual academics and the external world (May with Perry 2011). However, it is said that we now live in a post-Fordist knowledge-based economy, which has profound implications for universities acting at global and local scales. Universities have become implicated in local growth coalitions, not only as estate managers, but also as strategic actors, employers, or providers of evidence to inform policy (Perry and May 2010a). Yet tensions can and do emerge between the orientations of universities and the local contexts within which they are located (Pinheiro, Benneworth, and Jones 2013).

As universities become further intertwined with urban growth coalitions, we need to consider what the implications are for urban knowledge and systematically reevaluate the relationship between urban research and practice. This means reciprocal and *three-way* relationships between universities, urban researchers, and the city. Without a reflexive disposition, the danger is that the knowledge produced by the university loses its critical edge. It is no longer a question of what the city means for the researcher but also what research means for the city (May and Perry 2011a).

As problem and playground, the city has provided the academic with "a source of energy, of wonderful complex, intellectual problems and of non-academic intellectuals who have much to offer" (Bender 1998: 22). As such, "metropolitan academics ought not to work so hard at keeping the city at bay" (ibid.). Depending on the different disciplines and approaches of the researcher, the city has been both the object and location of study. Cities are crucibles to examine broader socio-economic, cultural, and political dynamics; a lens through which a route from the particular to the general can be traced. What we understand by the "city" is also variable—it is a concept constructed by different individuals in multiple ways; contested, complex, and infinitely reimaginable.

These issues are examined through the lens of the Centre for Sustainable Urban and Regional Futures (SURF) established in 2000 as an interdisciplinary urban research center at the University of Salford Manchester. The chapter is written from our particular points of view and positions within SURF—Tim as Lead Director since 2001 and Beth as researcher, Senior Research Fellow, and now Associate Director.[1] We draw on our experiences to discuss how the different actors within the urban field are motivated by diverse, and sometimes conflicting reasons to engage in knowledge-based development, related to complex and contradictory pressures from across academic and practice agendas. The mutual constitution of research and policy agendas is revealed, as well as a central tension between critique and co-option. We emphasize the need for a reflexive disposition in triangulating between the contexts, cultures, and consequences of urban knowledge production. In so doing, the chapter raises the need for more nuanced accounts of the dynamic interactions between research, policy, and practice worlds.

The chapter starts with a short history of SURF and the political–economic context of the city–region, emphasizing drivers and motivations for engaging in a knowledge-based urban agenda from different perspectives. Second, we highlight some of the tensions that can arise in engaged urban research practice and how we have sought to make sense of these experiences through theoretical developments and methodological reflections. Finally, the chapter reflects on recent changes in the urban research, policy, and practice landscape in the UK and internationally, such as Mistra Urban Futures, which offer the promise of a new approach to university–city–research relationships. The first two parts of the chapter focus on a ten-year period (2000–2010), with recent experiences discussed in the final section. The chapter is well grounded in a variety of academic literatures. However, as with all the chapters in this volume, we have sought to strike a tone that engages with a potentially wide readership. Consequently, the chapter acts as a summary guide to our work over this period of time, gleaned through a range of different research projects. Cross-references are made to our other writings that extend the theoretical and methodological arguments in particular directions.

Drivers, motivations, and rationalities for knowledge-based urban development

"Working for an Informed Future": the Centre for Sustainable Urban and Regional Futures

SURF was developed to generate understanding about how political, economic, social, technological, and environmental changes interact to affect urban and regional futures. Its original purpose was to promote interdisciplinary analysis of the sustainability of these changes and to disseminate knowledge to improve understanding and inform feasible actions for policy and practice. Drawing on endowments from two private sector companies in the northwest of England, as well as academic research funding from the UK's Economic and Social Research Council, SURF has drawn on multiple sources of funding, what we call "a mixed economy of research." This has involved funding from academic research councils, European framework programs, national and local governmental organizations, charitable or research foundations, and business organizations. A central prerequisite has been flexibility and capacity to bid for and subsequently deliver a wide variety of projects, deploying multiple methods and producing outputs for different audiences at multiple scales.

SURF has occupied different positions in its history on the continuum between research center, think-tank, and consultancy. Despite being largely self-financing, SURF has been driven by a clear vision and core ethos to guard against a fully commercialized approach. The SURF ethos (Box 5.1) was formulated to reflect a commitment to producing academically excellent research with high societal relevance. This was underpinned by the need to acknowledge the distinctiveness of academic knowledge (May and Perry 2006) as well as the responsibility of universities, as largely publicly funded organizations, towards society at large (Harloe and Perry 2004). It was therefore important to operate as a "critical friend" (Costa and Kallick 1993), actively interpreting and mediating between different values and motivations for urban development. The danger we were seeking to avoid was producing "policy-based evidence," described by Boden and Epstein (2006: 226) as resulting from the "capture and control" of the knowledge-producing process, such that evidence is produced to justify predetermined decisions. In this we grounded the research within literatures on social studies of science and technology, political science, geography, and urban studies.

Box 5.1: The SURF ethos

SURF is committed to the production and dissemination of knowledge that adds value through informing meaningful and desirable changes in our cities and regions. It combines the rigor of an academic research center with the client-oriented focus of a consultancy. It deploys knowledge to inform policies in ways that take the best of existing practices and builds upon them to support the achievement of sustainable urban and regional development.

SURF provides the space in which practitioners and policy-makers can reflect critically on the performance of policies. Key to this is the value placed on listening and learning processes that produce effective outcomes.

Tim May, Director, 2001

Implementing the SURF ethos and approach has meant working around core agendas and seeking funding sources that have enabled either a deepening of existing work or a broadening to new constituencies. A central example relates to the work around regional and urban science policy, which operated across the boundaries of academic and consultancy work, through commissions for local authorities and regional development agencies, alongside critical academic research (Perry 2012; Perry and May 2007, 2011). With few exceptions, non-academic contracts were only accepted on the basis that the work would further scholarly interests and contribute to what Tim termed, the "intellectual under-labor" of SURF. Multiple forms of representation, writing, and dissemination were undertaken to position the academics with different audiences, such as the briefing papers titled "SURF-Arup Framework for Urban Infrastructure" (SURF 2010a), "Active Intermediaries for Knowledge Exchange" (SURF 2010b), and "Integrated Visions for Knowledge Cities" (SURF 2011).

Within the work of SURF we were engaged in not only studying but also constructing potential pathways and responses to the higher education and urban agenda. We were mindful of the need to remain aware of how academics may become implicated as witnesses and agents in the development of possibilities for action; critical of those university strategic managers who unproblematically represent their institutions as the answer to urban prosperity (Perry 2012). A central methodological prerequisite has therefore been a reflexive orientation, to guard against bias and establish the validity of claims being made (May and Perry 2013a). At Tim's instigation, all members of SURF—from the directors to the administrative staff—presented and discussed their motivations, interests, and passions to the team, reflecting the desire to create the possibility for "intensive intellectual and affective fusion" (Bourdieu 2007: 19–20). A statement of values was produced and shared between the team.

The SURF approach was exemplified in management and governance structures within and outside the university. Until 2010 SURF was a cross-faculty center with staff located within different university schools, reporting directly to the Pro-Vice Chancellor of research through a management board. An external advisory board was also developed comprising leading regional and local figures from the public and private sector. Geographically, SURF was located away from the University of Salford main campus in central Manchester. Manchester and Salford are two of the ten local authorities comprising Greater Manchester, a metropolitan area of 2.6 million people in northwest England. Within the larger metropolitan area, Manchester and Salford represent a core of both economic activity and urban deprivation. Three of the four Greater Manchester universities—University of Salford Manchester, Manchester Metropolitan University, and University of Manchester—are within the two cities. The decision to locate SURF outside the University campus was a deliberate move to foster greater inter-university links, as well as represent the outward city–regional facing components of the work. A central location was appropriate to facilitate connections with national, regional, and local policy-makers who themselves were increasingly concerned with the relationship between research, policy, and practice.

Building knowledge cities in Greater Manchester

Throughout the late 1990s, and early 2000s a series of factors converged to create a fertile climate for sub-national mobilization in science, technology, and innovation. Critical policy factors which enabled the catalysis of regional science and innovation interventions included:

- The rising importance of the relationship between science and economic development and the importance of science exploitation in wealth creation.
- A shift to a more deliberate policy of concentration, in the context of discourses around the global "knowledge economy," economies of scale, and critical mass in light of concern with spatial proximity as a prerequisite for economic spillovers and scientific synergies.
- Increasing relationships between higher education institutions and regional actors, as a result of multiple factors, including national policy frameworks and the need to generate additional revenues.
- A failure to address the gap in productivity and prosperity between England's regions, with old industrial regions in the north of England deemed to be "underperforming" or "lagging" in the context of a national view that the performance of every region mattered in the national economic system.
- The establishment and increasing confidence of the Regional Development Agencies, coupled with high expectations of government and of regional policy in improving the economic performance of the English regions and leading to greater political devolution.
- An overall buoyant economic picture at the global and national level, despite competition from the rising economies of Brazil, Russia, India, and China and an increasing polarization between the haves and have-nots in economic, spatial, and social terms.

Developments in the wider northwest region and the Greater Manchester sub-region coevolved. In 2002 the northwest became the first region to establish a Science and Industry Council, led by the Regional Development Agency; and other English regions soon followed (Perry 2007). These brought university, industrial, and public sector agencies together to improve the innovation infrastructure and science exploitation in the region. Nationally, the Chancellor of the Exchequer announced the designation of six Science Cities in a 2004 speech. The 2005 Budget report linked the value of Science Cities to their potential to increase investment in science and research, to enhance the ability to exploit excellent science, and to compete in the global knowledge economy (May and Perry 2011b). Science Cities were announced for Manchester, Newcastle, York, Birmingham, Bristol, and Nottingham (see Box 5.2).

Box 5.2: Visions for the knowledge-based city

A definition of "Science Cities"	*The vision for Manchester: knowledge capital*
"Science cities" are those with strong science-based assets, such as a major university or centre for research excellence—which have particular potential to use these assets as the basis for generating business success. Developing science cities requires a range of complementary policies to address the specific needs of research and development, to support university–business collaboration and to influence a wider spectrum of factors that contribute to the growth of knowledge-intensive industries, such as skills, transport, finance and infrastructure. By bringing these factors together in a concentrated space, science cities can attract a critical mass of innovative businesses and become drivers of regional growth.	To create an internationally acclaimed 'Knowledge Capital' within the Greater Manchester conurbation, which will position Manchester, branded as the Knowledge Capital, at the heart of the Knowledge Economy, significantly contributing to the economic growth of the nation and the North West region leading to a healthier city/region with a vibrant, safe and attractive environment in which to live, work and play, for people of all ages, social and cultural backgrounds.
UK Government, 2005 Budget, Ch. 3, 'Meeting the Productivity Challenge'.	Manchester: 'Knowledge Capital', 2003.

Greater Manchester had, however, already developed its own vision for knowledge-based urban growth back in 2002, under the name of "Manchester: Knowledge Capital" (M:KC). M:KC was developed in 2002 as a result of discussions between the City of Manchester and the Work Foundation who produced a study on Manchester's potential to become an "ideopolis," building on such assets as enhanced cooperation between universities and the city, its concentration of knowledge workers, the presence of a medical school, and an international airport. The partnership brought together the ten local authorities, four universities, the regional development agency, key public agencies, the strategic health authority, and leading businesses to address four wide-ranging goals: (1) to stimulate and support business innovation from research, science, and knowledge; (2) to engage the people of Manchester through dialogue, debate, education, and employment; (3) to support the growth of a city–region environment which facilitates the championing and success of businesses; and (4) to try new ideas and new ways of living and working. "Manchester: Science City" became a sub-theme of M:KC, which also dovetailed with the regional science strategies.

M:KC was originally funded through local public bodies and the Regional Development Agency, the latter also supporting specific projects such as the merger between University of Manchester and University of Manchester Institute of Science and Technology in 2004. However, in 2007, a significant boost was given to the city–region when a partnership between the Regional Development Agency, Manchester City Council and the National Endowment for Science, Technology, and the Arts announced a £7m. Manchester Innovation Investment Fund (MIIF). Supported by the M:KC partnership, MIIF was established to deliver a "step change" in the innovative performance of the city through developing projects in Greater Manchester as a test-bed for new approaches. The MIIF was designed as an experiment in financing innovation and transforming the innovation ecosystem of a city–region.

Working together for an informed future?

As one of the primary urban research centers in the city–region, SURF played multiple roles in all of the above initiatives: constructing and constituting agendas, developing briefs, delivering work, and evaluating progress. Some of our projects were directly contracted by local partners, while others were funded by research councils or other academic funders. The expectations and anticipated outcomes therefore differed, in terms of providing knowledge for policy/practice or critical intellectual developments. The range of our projects is highlighted in Table 5.1.

SURF received national research funding from the Economic and Social Research Council (ESRC) for a study of the development of regional and local science policies; we carried out interviews with academics and senior managers funded by the Contact Partnership of Universities to support the development of M:KC, and carried out a feasibility study for a Greater Manchester Urban Knowledge Arena; we were commissioned by the Science Cities Policy Development Consortium to assist the Science Cities in representing their case to national government and share learning between them; and we designed and carried out a formative evaluation of the MIIF.

Some of our other projects developed knowledge and understanding that directly or indirectly sought to inform city–regional thinking, such as the transformational program with the Government Office North West, "From Information to Intelligence" and the ESRC Impact of Higher Education Institutions on Regional Economies. In the majority of cases, projects were underpinned by strong cross-organizational partnerships: for instance, EU funding was transferred from SURF to M:KC for their involvement in a Framework 7 Project on "Changing Behavior," while three projects involved partnerships with the University of Manchester and Manchester Metropolitan University.

Table 5.1 Selected SURF Projects in the Greater Manchester City–Region, 2000–2010

Project	*Dates*	*Type of project*	*SURF role*	*Partners*
"Making Science History": The Regionalization of Science Policy	2002–3	Academic; UK national research council	Independent researchers	None
Knowledge Capital: From Concept to Action	2003	Consultancy	Contracted researchers for the Contact Partnership of Greater Manchester Universities	None
Building Science Regions in the European Research Area: Governance in the Territorial Agora	2004–7	Academic; UK national research council	Independent researchers	University of Newcastle
Government Office North West (GONW): From Translation to Transformation	2006–7	Collaborative process and organizational development	Researchers as facilitators contracted by GONW	None
Renewable Energy Acceptance and Tools (CREATE Acceptance)	2006–8	European Framework project	Independent researchers	European partners
The Embedded University in the Science Economy: Capacities, Contexts and Expectations	2006	Academic Research Network Grant	Setting academic agendas	University of Manchester; University of Newcastle
The Greater Manchester Urban Knowledge Arena	2006–7	Consultancy	Contracted researchers for the Contact Partnership of Greater Manchester Universities	University of Manchester; Manchester Metropolitan University
Realizing the Potential of Science Cities: Evidence, Practice and Transferable Lessons	2006–7	Consultancy	Contracted researchers for Science Cities Policy Development Consortium	None
Manchester Innovation Ecosystem Analysis	2007	Consultancy	Contracted as independent expert by Manchester Enterprises	None

Table 5.1 (continued)

Project	*Dates*	*Type of project*	*SURF role*	*Partners*
Business and the Knowledge-Based Region	2007–8	Academic Research Council Impact Grant	Engaged researchers; facilitating knowledge exchange with business	British Business Chambers
Changing Behavior in Energy Demand	2008–10	European Framework program	Independent researchers	European partners plus Manchester Knowledge Capital
Urban Knowledge for Shaping Critical Infrastructure: New Models for City Regions	2008–9	ESRC Business Placements Fellows Scheme	Placement in global consultancy firm, Arup	Arup
Northern Way Research Program on Innovation and Symposium	2009–10	Discussion and debate series	Independent expert	University of Newcastle
Formative Evaluation of Manchester Innovation Investment Fund	2008–10	Consultancy	Formative evaluators; critical friend	University of Manchester, Regeneris consultants

In performing these roles we were both inside and outside formal policy processes, at times constituting agendas and at others evaluating them. Over the ten-year period from 2000–2010, urban research and policy agendas in the city–region entwined around the knowledge-based agenda, creating conditions for the flourishing of both the SURF center and Greater Manchester's multi-stakeholder partnership for city–regional innovation. For SURF we had tapped into a rich vein of work that could exemplify the ethos of working between research and practice, demonstrating impact, and sustaining intellectual agendas over time, in a context of multiple funding streams. For external organizations, academic partnerships offered credibility and legitimacy, as well as access to funding that might not otherwise be available. As we moved through the projects, our relationships to policy and practice became more interrelated, culminating in the formative evaluation of the Manchester Innovation Investment Fund, which saw us embedded in governance processes as critical friends, attending meetings, observing, and providing insights and intelligence in real time.

To a certain extent, this reflects the co-constitutive nature of urban research and policy. As English subnational governance shifted from regional to local levels, so too did research agendas. As discourses moved from sustainability to climate change and low carbon, so too have policy investments and research agendas. Yet this masks profound differences in motivations and rationalities for action. SURF's commitment to interrogating the sustainability of current political, economic, social, environmental, and technological change implies a critical but committed form of engagement. Underpinned by a reflexive disposition (May 2010), we were conscious of both our privileged position and the need to produce knowledge that could inform and seek to alter current conditions, while retaining its validity and robustness. Given changing conditions of knowledge production, the complexity of the urban, and the need for inter- and transdisciplinary working, we saw urban research–practice engagements as driven by factors both within the social scientific domain and outside. Central to all our projects was a concern with questions of inclusion/exclusion, redistribution, transparency, accountability and responsibility, drawing on a range of academic and non-academic perspectives. Our agenda was framed by the need for universities that were both places of expectation and spaces for reflection, critical of the narrow economic instrumentalism of the impact agenda (May and Perry 2006). In this, we were driven by the desire to question the taken-for-granted "necessities" of the policy world and by a meliorist belief that change is not only necessary, but possible (Koopman 2006; Stroud 2010).

This contrasted profoundly with the multiple orientations and expectations of urban stakeholders. Overall the national context for universities' engagement with their cities and regions in the 2000s was shaped by a dominant discourse of the knowledge-based economy and its multi-scalar manifestations. Academics such as Manuel Castells had described universities as "being to the knowledge economy what coal-mines were to the industrial revolution" (Castells and Hall 1994). Others, coming from business schools, drew on experiences of emblematic success stories—such as Silicon Valley and the development of the Massachusetts Institute of Technology (MIT)—to argue for the centrality of universities within urban economic growth coalitions, in partnership with industry and government (Etzkowitz 2002). Central among these were ideas concerning localized cluster development thanks to Michael Porter (1990, 2003) and Richard Florida's "creative class" (2002), representing a particular kind of vision for future cities—one based on the economic and creative potential of particular kinds of knowledge workers, rather than the emancipatory or transformational possibilities of knowledge.

Not all urban stakeholders held the same positions. While the national Science Cities initiative was driven by globally oriented, nationally focused concerns with scientific–technological development and economic growth, M:KC appeared initially to reflect an alternative approach around a more holistic, knowledge-based vision of how universities and other partners could engage for the benefit of different groups in the city. Importantly, and in contrast to other initiatives, it was run as a partnership, an intermediary body accountable to the Regional Development Agency operating on behalf of all the city–regional stakeholders. The vision (Box 5.2) highlighted concern for the health, safety, and quality of life of different citizens; however, in terms of outcomes, the investment framework (2007) noted that "the outcomes of the project will be higher rates of wealth creation and economic performance in the Manchester city–region, as a result of a more complete innovation ecosystem and further subsequent development of the knowledge economy." As we have explored elsewhere in more depth (Perry 2008), what emerged as the dominant narrative was a neo-liberal entrepreneurial model of urban development, which equated sustainable economic development to growth, instead of a more integrated approach. Activity was driven by the need to create economic prosperity and redress regional disparities in growth—neither being clearly linked to either ecological or social concerns.

Despite an apparent coincidence of interests, the drivers, motivations, and rationalities for engagement in knowledge-based urban development varied between SURF and mainstream urban stakeholders. What emerged was a difference between visions of a knowledge economy (for the few) and a knowledge society (for the many), the former underpinned by the need for economic growth as a precondition for redistribution, and the second linked to the idea that we need new ways of thinking, learning, and doing in our urban environments. A tension was evident between the values and motivations (the "normative" position) of SURF, our ways of knowing (our "epistemological" position) and those of key actors in the city–region, resulting in a normative–epistemological clash.

Tensions in and possibilities for the urban research–practice relationship

Devilish dichotomies between global and local, excellence and relevance

These diverse drivers, motivations, and rationalities translated into different conceptualizations of the role of universities and knowledge in urban development. Over time, city–regional projects involved the Greater Manchester universities in various ways. Initially all universities had engaged with the Regional Science Council through the North West Universities Association, however it quickly became apparent that the agenda was being set by and oriented towards the research-intensive universities and a very particular science-based agenda of economic growth. Within Greater Manchester, the universities continued to take some collective action through the Contact Partnership. This body brought together senior representatives—usually Pro-Vice Chancellors/Deputy Vice Chancellors—of the Universities of Manchester, Manchester Metropolitan University, and Salford University. They commissioned two key reports in the early 2000s to examine their contributions to the developing knowledge-based urban agenda. The first report, which we named "Knowledge: Capital. From Concept to Action" (2003) drew on our interviews and desk-based reviews to illustrate the different ways in which universities saw their roles—some emphasized business as usual but with an urban slant, while others saw the potentials for a more transformative set of outcomes. What was apparent was the vast diversity across and between institutions and individuals of what "knowledge capital" could and might mean, accompanied by strong variations in the willingness of individuals at different scales of action to engage in locally relevant work (May and Perry 2006). A clear issue was uncovered in terms of what universities could and should contribute to M:KC, but also how and by whom it should be led.

In 2006 a further piece of work was commissioned by the Contact Partnership, this time for a project bringing together research staff from each university to examine the creation of a Greater Manchester Urban Knowledge Arena (May 2011). The idea originated not only in the early Knowledge Capital work, but also through a multiplicity of SURF's previous initiatives including proposals for an urban foresight laboratory (2001), a center of excellence for regeneration skills and practices (2001), and a scoping study for a UK national research program on the impact of universities on their regions and localities (2006). The project sought first to map existing institutional strengths and collaborations in three fields: sustainability and the built environment; health; and creative and cultural industries. The intention was then to compare these initial areas of existing research against city–regional priorities to see whether, and if so where and how, synergies could be fostered to produce a shared city–regional research strategy.

The project never reached beyond the first phase for four key reasons. First it was clear that there was not—and could not be—a single point of contact or access within the universities who could represent the entire range of research carried out within their walls. Second, while collaborations were apparent between the universities in the city–region, global networks and disciplinary communities were stronger determinants of collaborative behaviors than proximity or local relevance. Academic practices were bound neither by their institutions, nor their spaces, working outside of the city–region as much as within it. Rarer still was research on, for, or with Greater Manchester—even in clearly locally relevant fields such as sustainability or the built environment. This highlighted a third issue, that of the gap between levels and hierarchies within institutions, with a disconnection between the visibility of the city to strategic managers and to individual academics. Most importantly, the project tested the limits to institutional collaboration. Universities were not able to agree on their strengths and areas for collaboration, nor on how their strengths would be represented externally, resulting in a retreat into competition and a jostling for position in local knowledge hierarchies (Perry 2012).

For so long as M:KC was an umbrella organization with few resources beyond core costs, all partners could unproblematically support it. However, as the Manchester Innovation Investment Fund came on stream, the prospect of funding, combined with a rapidly changing context, marked the beginning of the end to formal institutional collaborations. Universities were positioned less as partners and more as beneficiaries of the MIIF projects, such as in the development of Manchester Masters, the Manchester: Integrating Medicine and Innovative Technology (MIMIT) initiative, the Smart City Futures conference, or investments in the Oxford Road Corridor. The overall impact of the MIIF was limited by the complexity of multi-arena partnerships; the difficulties of mixed funding streams; differences in orientation and values; turf wars over who "owned" innovation, and changing national and economic contexts.

Throughout all of these initiatives, the impact on individual academic practices vis-à-vis urban areas was limited. Universities engaged as strategic actors, but questions over academic knowledge production or processes were not addressed. This resulted in an acquisition- and product-based model and missed the potential for more holistic approaches to urban engagement to emerge (see Table 5.2). Fundamentally, knowledge was seen as an ingredient to be acquired, underpinned by linear models of knowledge transfer, rather than new processes of knowledge production (see Chapter 1).

This was reflected in how we ourselves were positioned. Over time we became gradually more embedded in city–regional processes, culminating in the formative evaluation of the Manchester Innovation Investment Fund. Our role, alongside the University of Manchester's Institute of Innovation Research, was not to act as consultants, but as "critical friends," providing real-time feedback and lessons that could inform subsequent actions, rather than producing an end-of-project summative evaluation. The mode of working was developed at our instigation—indeed, we wrote the brief—drawing on our critique of academic engagement and linear models of knowledge transfer as opposed to those concerning "exchange." Our aspiration, shared by one of the key organizations involved, was to bring endogenous and referential concerns, excellence, and relevance together in a novel approach to inform practice not as a by-product or end product, but as an intended, collaborative outcome (May with Perry 2011). We can see this as embodying a frustration with existing academic practices and a desire for a more critical, emancipatory form of research (see McCabe and Holmes 2009).

Over time, however, the normative and epistemological differences between our visions for knowledge cities and those of city–regional stakeholders became more apparent. Deemed "too accurate to see the light of day," our findings began to shine lights in places the funders were uncomfortable with. Collaboration quickly turned into a choice between co-option in support of existing agendas or critique. A central tension was the expectation from funders that academics should provide "answers" or "solutions" to public policy issues, while we argued for a more relational and joint sense of responsibility for addressing contemporary challenges. Our refusal to accept the isolating tactics behind an outsourcing of responsibility—or to capitulate to the demand for bite-size positive success stories—led ultimately to the termination of the evaluation after two, instead of three stages of work. Ultimately, the idea of "critical friends" gave way to a more traditional relationship of funder–client–consultant, expected to deliver set outputs rather than inform

Table 5.2 Types of knowledge-based urban development

	Influences	*Responses*	*Role of city*	*Role of university*
Acquisition driven	Urban growth theory and urban entrepreneurialism	Emphasis on the ingredients for success to be acquired; "what" rather than the "how" or "why."	Emphasis on branding and position. Marketing role for cities.	Universities as one among many partners. Strategic institutional actors in urban growth coalitions.
Product driven	Economics and innovation studies	Knowledge as a product to be exploited by "end-users."	A passive, facilitative role in terms of creating an environment for innovation.	Universities as providers of knowledge; emphasis on business–university interactions.
Process driven	Social studies of science and technology	Knowledge as a process; new modes of knowledge production, ways of working, user engagement and knowledge exchange.	An active role leading to varied political, economic, and social benefits.	Co-produced knowledge and multiple engagements. Fluid and diverse interactions, difficult to map. Universities providing strategic intelligence for action.

Source: Perry (2008).

processes of learning and reflection. Consequently, we then created new spaces to reflect on the work, in part through removing ourselves from our environment and presenting at conferences, not on the work itself, but on the process (Knoblauch 2004: 357). This ongoing reflexive re-analysis of our roles and the relationship between creativity and critique in urban governance and policy processes acted not only as "confession, catharsis and cure" (Pillow 2003), but also to inform our current practices.

Across the range of our work, we highlighted that expectations and aspirations for change outweighed capacities to deliver; that global positioning appeared to drive activities more than local relevance; that narrow interpretations of innovation predominated over more holistic understandings of knowledge; that institutions competed rather than collaborated; and that experts and elites designed and determined policies without taking lay and non-professionalized knowledges or communities into account. As John Maynard Keynes is quoted as stating: "there is nothing a government hates more than to be well-informed; for it makes the process of arriving at decisions much more complicated and difficult" (Skidelsky 1992).

Predominantly the innovation agenda, like so many areas of urban public policy, was characterized by a "missing middle": many policy-makers believed they knew what was needed, although this was not reflective of the multiplicity of stakeholders in the urban realm. Policy-makers lacked knowledge on how to move beyond a linear model of innovation and science-based growth towards more sustainable, knowledge-based urban futures. A series of tensions—or devilish dichotomies—emerged between global and local scales of action, excellence, and relevance and the discourses of the knowledge economy or knowledge society (Perry and May 2010b).

Constructing possibilities for action: populating the missing middle

Through our work we identified a "missing middle" in the research–policy–practice relationship, characterized by a mismatch in understanding the dynamics of contexts of knowledge production and knowledge reception. For academics there was a gap between the expectations placed on knowledge (excellence and relevance), incentives, and evaluation criteria. For policy-makers, receiving information on what to do is often meaningless without understanding how knowledge can become embedded, understood, and utilized in different organizations. We therefore sought to populate the missing middle by suggesting alternative narratives and practices for knowledge-based urban development. We argued in pamphlets and academic articles that existing knowledges and accepted wisdoms need to be unbundled and critically assessed to better understand how dominant models are developed and transferred and what the potential implications are for cities and city–regions in different contexts. We concluded that an alternative rebundling of a wider set of knowledges at the urban level is needed, to bring different social interests and visions together for a more sustainable urbanism to develop (see Table 5.3).

Table 5.3 Dominant and alternative discourses for knowledge-based urban development

	Dominant responses	*Alternative responses*
Objectives	Econo-centric	Varied: triple line of sustainability
Measurements	Tangible	Intangible
Scales	Global excellence	Glocal "excellent relevance" and "relevant excellence"
Processes	Linear, products, supply–demand, push–pull models	Ecosystems, networks and flows
Knowledges	Narrow; disciplinary; sectoral; codified	Broad; interdisciplinary; cross-sectoral; tacit
Mechanisms	Technological, mechanistic solutions	Multiple interventions and mechanisms
Learning	Transferable models	Context-sensitive approaches
Social interests	Elites and experts: corporate, governments, major institutions	Wide stakeholders, potential beneficiaries and participants

Source: Perry, May, Marvin and Hodson (2013).

Strong epistemic hierarchies currently exist, in which knowledge for understanding, such as that generated traditionally through the social sciences, is valued less than the (assumed) commercially exploitable knowledge of the physical sciences. We require differentiated knowledge about, for, and in cities to rethink how context-sensitive approaches can be developed to address the economic and ecological challenges of the twenty-first century. Such knowledge needs to be interdisciplinary and multi-sectoral, have global resonance and local relevance, and be co-produced to ensure the effective exchange of knowledge between producers, users, and consumers (Andersen and Atkinson 2013).

The capacities and capabilities of cities to achieve such ambitions are limited by tensions in the international political economies of knowledge production and use for sustainability. An uncritical embrace of neoliberal competitive discourses results in polarized discourses between excellence and relevance, the global and the local, the economic and the social, and the commodified and the intangible (Perry and May 2006). Frameworks for action—whether for funding and evaluating research or for urban policy development—embody mixed messages and frequently leave cities themselves responsible to mediate between competing claims. The reasons for this lie in the ways in which particular forms of knowledge (May and Perry 2013b) have been attributed value because they result in a product, rather than because they have been constituted through a process. In the latter case, success is more intangible and not amenable to crude linear measures of output that stifle innovation.

Innovation in its broadest sense is not one-way—it requires greater dialogue and partnership between the knowledge base, public sector, and business communities. There are no simple solutions or quick fixes. To move away from such a state of affairs requires a mental unshackling from the perceived need to search for the holy grail—the model, the software, the platform, the form of organization—that can deliver sustainable knowledge-based urban growth. In its place must be a concern for ways of working in, thinking about, and engaging with cities—for national and sub-national actors across sectors alike. The focus needs to be on the creation of new agendas that are sensitive to different places, not simple solutions proposed at national level with an indifference to the contexts of their implementation.

In SURF we sought to construct this alternative agenda through our engagement with policy processes, by seeking to influence existing initiatives and proposing new ones. These included a proposal for developing a Center of Excellence in Regeneration in the North West and an Urban Foresight Lab to anticipate, evaluate, and shape sub-regional futures in 2001, as well as the Greater Manchester Urban Knowledge Arena project outlined above. We sought to capture the implications of our ten-year engagements through the idea of "active intermediation" between different cultures of knowledge production and reception. Active intermediation was conceived not as a solution or model to be implemented, but as a set of practices in the interstitial spaces between research and practice. As such, it represents the active and constant "agonism" (Mouffe 2000) of engaged urban research: there is no state of resolution, rather a set of practices that inform the possibility of producing excellent–relevant knowledge. This requires intensive work at understanding the boundaries, limits, and consequences of collaborative working. Fundamentally, however, our efforts were undermined by the simple fact that we were funded to support and implement the status quo, rather than to challenge or improve it.

2010+: towards urban co-production

The end of the urban innovation paradigm?

By the end of the 2010s the urban innovation bubble was under stress. Global, national, and local issues intersected to cast doubt over the long-term viability of efforts to improve the relationship between research and practice. These included the credit crunch of 2008 and the ensuing financial crisis and recession; a changing national policy context for higher education with the introduction of tuition fees for students building on the Dearing Report; a continued emphasis on world-class science and innovation within the context of an increased emphasis on growth; a shifting emphasis from innovation to the role of cities in addressing climate change; and increasing competition between places for location-specific assets. Universities had not catalyzed around a city–regional knowledge-based agenda; while the public sector had taken the initial risks, expected private sector investment had not materialized.

Furthermore, a spatial bifurcation of efforts for place-based development was emerging between Salford and Manchester, the two cities comprising the city–regional core. In particular, the decision to move the British Broadcasting Corporation (BBC) to Salford Quays contributed to two parallel, but largely disconnected sets of initiatives for university–urban partnerships. Corridor Manchester, a project with multiple roots including funding through the MIIF, brought together the City of Manchester, the Manchester-based universities, and other public/private stakeholders. On the other side of the river Irwell, Salford University was consolidating developments around Media City and the creative and cultural regeneration of the docks. Both projects are mentioned in Greater Manchester strategy documents as key building blocks for the knowledge economy. However, the institutional architectures that supported a collaborative approach were not to last. The Contact Partnership was disbanded; M:KC has also vanished, after a protracted debate about its future which saw it temporarily rehoused in Manchester Metropolitan University before being permanently wound up. The MIIF ran out of funding; while some projects continued, such as MIMIT or Manchester FabLab, the collaborative infrastructure and partnerships to support innovation were replaced by occasional discussions on LinkedIn.

With hindsight it is clear that the initiatives themselves were neither designed nor run with a view to an integrated or sustainable approach to the development of the city–region. Discourses related to both sustainable economic growth and sustainable urban development could be heard from different quarters throughout the decade, and Greater Manchester's initial imaginations of a knowledge-based city were broadly reflective of a "triple bottom line" approach. The vision for M:KC included new ways of living and working, health, quality of life, and attractive urban spaces. One proposal in the original report we wrote for M:KC called for "sustainable city-thinking" through a process of looking for overlaps, synergies, and potentials for collaborative working across the Greater Manchester universities for city–regional benefit. Both our independent reports for the Contact Partnership flagged up sustainability as a critical theme, as did the work conducted for the Science Cities Policy Development Consortium. Manchester: Science City did in fact include a working group around design for sustainability and Manchester is My Planet was taken on as a key program within M:KC. However, economic growth was the primary driver and alternative discourses were largely marginalized. Even where environmental and ecological projects began to be funded through the MIIF, this reflected the rise of an economically driven, low-carbon and green-growth agenda and the positioning of Greater Manchester to be a first-mover or tested for national policy initiatives, such as the Low Carbon Economic Area, rather than an embedded concern for sustainable urban development.

From this point of view, initiatives in the 2000s were not sustainable—they did not last. Several contributory factors can be identified. External factors included austerity, financial recession, and changes in the higher education landscape. Funding cuts were officially blamed, leading to a major restructuring of the Manchester "family" of organizations, with the website noting that "the Manchester: Knowledge Capital partnership work will continue to be delivered within the Manchester Family of Organizations, but in the current financial climate, the Board has had to recognize that an independent organization is no longer viable." However, a series of internal pressures and limits are also explanatory factors, including the way in which governing bodies interacted and the limits to creating intermediary, truly collaborative partnerships that coalesced around common values (Perry 2011). Limited to strategic relationships with universities, developments largely bypassed individual academics with little focus on producing new knowledge, or recombining existing knowledges, for city–regional benefit. Models of innovation were largely linear and individualized, constituted by a winner-takes-all economic growth paradigm. Importantly, the fizzling out of the city–regional innovation excitement in the 2000s can be seen as a triumph of symbolic and aspirational politics over tangible initiatives, resulting in no clear approach to measuring and assessing outputs or, more importantly, outcomes for different groups. A knowledge economy had emerged in name, but not in substance or content.

Strengthening engaged urban research agendas

While policy-based initiatives were halting, academic agendas constituted around learning from and addressing the challenges of the innovation agenda in the 2000s were being consolidated. Through our previous work, we had argued that a singular challenge was to incentivize academic engagement in applied and co-produced urban research, when funding initiatives and evaluation criteria tended to reward and reinforce traditional forms of academic scholarship, prioritizing international reputation and publication in peer-reviewed journals. By the end of the 2000s, UK schemes were beginning to emerge, building on previously more ad hoc approaches to develop knowledge exchange activities within academic research projects. The Cross-UK Research Council "Connected Communities" program is one example, with an explicit aim to engage with communities at all stages of the research to co-produce research through new methods and tools, drawing particularly on participatory arts, digital tools, or techniques such as storytelling. SURF and a consortium of UK universities were successful in a collaborative bid to the Connected Communities program for a project to investigate the effectiveness of different mechanisms to connect communities into the creative urban economy. This was inspired by the all-too-evident disparities between the economic and social fortunes of deprived traditional urban areas and emerging creative quarters such as Media City (2012–2016). Over the four years of the project, we will trial, document, and reflect on a range of methods, including co-creating community snapshots of cultural activities, working with practitioner-researchers, recruiting participant diary-keepers, and establishing experimental community-led cultural commissioning panels in deprived urban neighborhoods, alongside participatory evaluation methods.

A strong global response to the challenges inherent in the research–practice relationship identified through our work can be seen in the launch in 2010 of the Swedish-led Mistra Urban Futures Center. Mistra Urban Futures is conceived as a global center for sustainable cities, with partners in Gothenburg, Cape Town, Kisumu, and Greater Manchester—core cases around which this book is based. In partnership with European partners, SURF was involved in both the feasibility work and design of Mistra Urban Futures Centre, drawing on our experiences, writings, and analyses. Importantly, Mistra Urban Futures Centre has been designed as an experiment in financing and constructing excellent–relevant research that is both globally and locally applicable and draws on the knowledge and skills of different academic and nonacademic knowledges. The distinctiveness of the center, still in its first phase of operation, has been in supporting co-produced local projects, while creating fora for international exchange and learning.

For SURF the initiative provides the opportunity to act on our own critique, to be engaged in the construction of new possibilities for action across the city–region. Like Gothenburg, Cape Town, and Kisumu, we have sought to establish a *local interaction platform* to embody new ways of working with academic, public, private, and voluntary partners. This has involved learning-by-doing about co-produced urban research, through a series of methods and tools (Table 5.4).

Table 5.4 Examples of co-producing research in the Greater Manchester Local Interaction Platform

Method	*Application*
Interactive research exhibition	Sustainable Stories exhibition, co-funded by Mistra Urban Futures and Economic and Social Research Council, involving the co-production of content between visitors and visual artists.
Policy-researchers	Working with policy-makers in the Greater Manchester Low Carbon Hub as researchers, co-producing challenges and options to inform policy.
Community researchers	Gathering sustainable stories from the ground with community researchers, working with a local development trust.
Dual leadership	Developing the online digital Platform, www.ontheplatform.org.uk with sustainable communications firm, Creative Concern.
Dual research teams	Working in a dual research team on realizing the potential of community assets with Greater Manchester Center for Voluntary Organizations.
Practitioner-researchers	Co-analysis and writing with practitioners to lend weight to their knowledge and expertise. Trialing method with Vincent Walsh, Director of the Biospheric Foundation.

The launch of Mistra Urban Futures coincided with a series of far-reaching changes in the political, economic and social climate in the UK. In 2010 a new Conservative–Liberal coalition government was elected, quickly followed by a deep financial recession and a far-reaching program of public sector reform. In Greater Manchester, this resulted in sweeping reform in regional and local governance structures. As a result, there was little continuity in personnel, unclear roles and responsibilities of key agencies and diminishing public sector funding and resource.

These issues shaped the context for creating a local interaction platform in Greater Manchester. Policy processes were complex and opaque, with new structures and policies being layered on top of old. Knowledges, skills and expertise were widely distributed and attempts to involve stakeholders and different community groups in formal governing arrangements had been partial. Furthermore, city–regional attempts to institutionalize relationships between universities and public sector bodies had been unsuccessful. Three challenges related to: addressing fragmentation and institutional confusion; interrogating different visions and pathways for sustainable urban development and reshaping the relationship between research and practice through the concept of co-production.

In response to these challenges and drawing on SURF's previous work, the Greater Manchester Local Interaction Platform (GMLIP) was developed as an "intermediary" space to populate the missing middle, integrating existing and developing new knowledge to inform sustainable urban futures. The aim is to increase capacity for sustainable urban futures through developing connections between disparate policy communities and often atomized community initiatives in the spirit of enhancing learning and flows of knowledge. Specifically, through co-developing a series of flagship projects with key city–regional anchor partners, the GM LIP has developed as a platform to raise the *visibility* of alternative forms of sustainable urbanism, give *voice* to those alternatives and contribute to the articulation of different *values* for city–regional futures.

These projects are, to a certain extent, all indicative of the researcher initiated mode which underpins this book (see Chapter 1). Yet the projects respond to different drivers and the roles and relationships of each actor vary across our work: in who sets the agenda, suggests the questions, does the work and analyses the results. Co-production does not mean that the boundaries between areas of knowledge collapse; rather it calls our attention to who is doing what and why in the knowledge exchange process. Our project with the Greater Manchester Low Carbon Hub, on challenges and pathways for addressing gaps in governance, knowledge and policy, was clearly researcher initiated and has involved practitioner-researchers not in the design of the work, but in its execution, management and analysis. Our action-research project to develop a digital portal for sustainability does not stem from a preconceived set of research questions, but rather responds to issues emerging from business, state and civil society and is steered by non-academic partners, Creative Concern. These are just two examples. They also point to an interesting set of questions over power and control and how this shifts over time.

Outcomes may be unsurprising or unexpected—suggesting that evaluations that assess co-production according to anticipated predefined objectives may be doomed to failure.

Much of this work is still ongoing, yet some early lessons have emerged. Co-production is not a panacea and requires careful management of expectations from all parties. While co-production offers a potential way to circumvent narrow funder–client–consultant relationships, sensitivity to different agendas and motivations remains critical. A university-led structure offers the opportunity to balance between critique and co-option, bringing about a different set of power relations, with academics commissioning from external stakeholders. However, the requirement from Mistra Urban Futures for co-funding can also become a complicating factor, potentially minimizing the possibilities for constructive critique. Here we must question what is distinctive about the knowledge produced by the university in terms of its ability to transcend the short-termism of policy imperatives. Co-production without a clear set of values runs the risk of muddying, rather than clarifying knowledge on how to create more sustainable forms of urbanism.

Summary

In this chapter we have taken a reflexive tour through our experiences in working for and with urban stakeholders to improve the research–practice relationship in the 2000s. Co-production highlights the need for a reflexive disposition to guard against co-option and protect the role of critique (May and Perry 2013c). The history of the SURF Center—in relation to the university, our urban and regional environment, funding regimes, and policy frameworks for research—has shaped the "degrees of reflexivity" we have been able to exhibit over the passage of time (Mauthner and Doucet, 2003). Since 2010 SURF has gone from a cross-faculty research center to one "normalized" within a single school. Notions of relevance, impact, interdisciplinarity, collaborative working, and the need to be outward-facing have collided with systemic organizational turbulence and restructuring, the latter brought about as a result of changes in the UK higher education landscape.

Not only our characters and commitments, but the contexts and cultures in which we have worked have shaped the "multiple reflexivities" we have practiced (see Lynch 2000; Mruck and Mey 2007). Our common predispositions to be reflexive have been mediated through our variable positions and career trajectories, through shared and individual relationships to SURF, the institution, and each other. Through our upbringings, histories, and biographies, we each brought with us multiple "ghosts" (Doucet 2008). Collectively manifesting and exploring these was a critical element in creating a culture in which reflexivity could be a relational and collaborative, rather than self-referential and individualized exercise.

We have highlighted how the context for engaged urban research practice has been shaped by a neoliberal economic agenda and narrow, econo-centric, and linear interpretation of the impact agenda. Partnerships have been forged, not around sustainable ways of living and working for the twenty-first century and beyond, but through business-as-usual public–private partnerships focused on economic growth. Broader concerns around civic engagement have been largely missing.

SURF has been both constituted by and constitutive of social relations around knowledge exchange and innovation. Yet normative and epistemological positions have diverged. It is for this reason that the current emphasis in funding schemes on co-production offers so much promise. Sustainability, while ontologically contentious, is perhaps a more hopeful starting point for revisiting urban research–practice relationships than innovation, technology transfer, and economic growth. The ten-year period 2000–2010 is characterized as one dominated more by competition than collaboration, by a growth-centric paradigm rather than the search for a more sustainable urban future. It is in this context that the Mistra Urban Futures initiative, as a bold experiment in urban research, finds its distinctiveness. The "conditions of possibility for critical urban research have changed in the early 21st century" (Brenner 2009); new research practices are needed that are both constitutive of and constituted by the promise, problems, and potential of the urban.

Acknowledgements

We acknowledge all of the funders that have supported SURF projects between 2000 and 2010 on which this chapter is based—specifically the Economic and Social Research Council, the Contact Partnership of Universities, the North West Development Agency, the National Endowment for Science Technology and the Arts, and Manchester City Council. The analysis here is our own.

We would specifically like to thank Mistra Urban Futures for funding the production of this chapter and the opportunity to analyze and reflect on our experiences. We also draw on ongoing reflections from the Arts and Humanities "Connected Communities" programme for our project on "Cultural Intermediation in the Creative Urban Economy".

Note

1 SURF started as a cross faculty Centre with Alan Harding and Simon Marvin as co-directors and Ian Cooper, who was appointed part-time by the University. Tim was seconded to SURF in 2001 and was asked to be the Lead Director. He held that role for four years and then Simon took over and he and Tim then co-directed SURF until Simon's departure for Durham at the end of 2010.

References

Andersen, H. T. and Atkinson, R. (eds) (2013) *The production and use of urban knowledge: European Experiences*. Dordrecht: Springer.

Bender, T. (1998) 'Universities and the city: scholarship, local life and the necessity of worldliness', in H. van der Wusten (ed.), *The Urban University and its Identity.* Dordrecht: Kluwer, pp. 17–29.

Boden, R. and Epstein, D. (2006) 'Managing the research imagination? Globalization and research in higher education', *Globalization, Societies and Education*, 4(2): 223–36.

Bourdieu, P. (2007) *Sketch for a Self Analysis.* Translated by R. Nice. Originally Published in 2004 as *Esquisse pour une auto-analyse.* Cambridge: Polity.

Brenner, N. (2009) 'What is critical urban theory?', *City*, 13(2–3): 198–207.

Castells, M. and Hall, P. (1994) *Technopoles of the world.* London: Routledge.

Costa, A. and Kallick, B. (1993) 'Through the lens of a critical friend', *Educational Leadership*, 51(2): 49–51.

Doucet, A. (2008) 'From her side of the gossamer wall(s): Reflexivity and relational knowing', *Qualitative Sociology*, 31: 73–87.

Etzkowitz, H. (2002) *MIT and the Rise of Entrepreneurial Science.* London: Routledge.

Florida, R. (2002) *The Rise of the Creative Class and how it's Transforming Work, Leisure, Community and Everyday Life.* New York: Basic Books.

Harloe, M. and Perry, B. (2004) 'Universities, localities and regional development: The emergence of the "Mode 2" university?', *International Journal of Urban and Regional Research*, 28(1): 212–23.

Knoblauch, H. (2004) 'The future prospects of qualitative research', in A. Flick, E. von Kardoff and I. Steinke (eds), *A Companion to Qualitative Research.* London: Sage.

Koopman, C. (2006) 'Pragmatism as a philosophy of hope: Emerson, James, Dewey, Rorty', *Journal of Speculative Philosophy*, 20(2): 106–16.

Lynch, M. (2000) 'Against reflexivity as an academic virtue and source of privileged knowledge', *Theory, Culture and Society*, 17(3): 26–54.

McCabe, J. and Holmes, D. (2009) 'Reflexivity, critical qualitative research & emancipation: a Foucauldian perspective', *Journal of Advanced Nursing*, 65(7): 1518–26.

Mauthner, N. S. and Doucet, A. (2003) 'Reflexive accounts and accounts of reflexivity in qualitative data analysis', *Sociology*, 37(3): 413–31.

May, T. (2010) 'Contours in reflexivity: Commitment, criteria and change', *Methodological Innovations Online*, 5(1): 24–44.

——(2011) 'Urban knowledge arenas: Dynamics, tensions and potentials', *International Journal of Knowledge Based Development*, 2(2): 132–47.

May, T. and Perry, B. (eds) (2006) 'Universities in the knowledge economy: Places of expectation/spaces for reflection?', *Special Edition. Social Epistemology*, 20(3–4).

——(2011a) 'Urban research in the knowledge economy: Content, context and outlook', *Built Environment*, 37(3): 352–68.

——(2011b) 'Contours and conflicts in scale: Science, knowledge and urban development', *Local Economy*, 26(8): 715–20.

——(2013a) 'Reflexivity and the practice of qualitative research', in U. Flick (ed.), *SAGE Handbook of Qualitative Data Analysis.* London: Sage.

——(2013b) 'Translation, insulation and mediation: universities in an age of ambivalence', in P. Benneworth (ed.), *Universities Engagement with Socially Excluded Communities.* Dordrecht: Springer.

——(2013c) 'Universities, reflexivity and critique: Uneasy parallels in practice', *Policy Futures in Education*, 11(5): 505–14.

May, T. with Perry, B. (2011) *Social Research and Reflexivity: Content, Consequences and Context.* London: Sage.

Mouffe, C. (2000) *The Democratic Paradox*. London: Verso.

Mruck, K. and Mey, G. (2007) 'Grounded theory and reflexivity', in A. Bryant and K. Charmaz (eds), *The SAGE Handbook of Grounded Theory*. London: Sage.

Perry, B. (2007) 'The multi-level governance of science policy in England', *Regional Studies*, 41(8): 1051–67.

——(2008) 'Academic knowledge and urban development: theory, policy and practice', in T. Yigitcanlar, K. Velibeyoglu and S. Baum (eds), *Knowledge-based Urban Development. Planning and Applications in the Information Era*. New York: Information Science Reference.

——(2011) 'Universities and cities: Governance, institutions and mediation', *Built Environment*, 37(3): 245–60.

——(2012) 'Science frictions and fictions in the North West of England: Excellence, relevance and the university', in R. Pinheiro, P. Benneworth, and G. Jones (eds), *Universities and Regional Development: A Critical Assessment of Tensions and Contradictions*. London: Routledge.

Perry, B. and May, T. (2006) 'Excellence, relevance and the university: The "Missing Middle" in socio-economic engagement', *Journal of Higher Education in Africa*, 4(3): 69–92.

——(eds) (2007) 'Governance, science policy and regions: special edition', *Regional Studies*, 41(8).

——(2010a) 'Building knowledge cities: The roles of universities', special edition, *Built Environment*, 37: 3.

——(2010b) 'Urban knowledge exchange: Devilish dichotomies and active intermediation', *International Journal of Knowledge-Based Development*, 1(1): 6–24.

——(2011) 'Case Studies', in T. May, *Social Research: Issues, Methods and Process*. 4th edn. Maidenhead: Open University Press/McGraw-Hill.

Pillow, W. (2003) 'Confession, catharsis or cure? Rethinking the uses of reflexivity as methodological power in qualitative research', *Qualitative Studies in Education*, 16(2): 175–96.

Pinheiro, R., Benneworth, P. and Jones, G. (eds) (2013) *Universities and Regional Development: A Critical Assessment of Tensions and Contradictions*. London: Routledge.

Porter, M. (1990) *The Competitive Advantage of Nations*. New York: Free Press.

——(2003) 'The economic performance of regions', *Regional Studies*, 37(6–7): 549–78.

Skidelsky, R. (1992) *John Maynard Keynes; A Biography. Vol. 2: the economist as savior, 1920–1937*. London: Macmillan.

Stroud, S. (2010) 'What does pragmatic meliorism mean for rhetoric?', *Western Journal of Communication*, 74(1): 43–60.

SURF (2010a) 'The SURF–Arup framework for urban infrastructural development', *SURF 10th Anniversary Pamphlet Series*, No. 2.

——(2010b) 'Active intermediaries for effective knowledge exchange: Populating the "Missing Middle"', *SURF 10th Anniversary Pamphlet Series*, No. 1.

——(2011) 'Integrated visions for knowledge cities', *SURF 10th Anniversary Pamphlet Series*, No. 3.

6 Multifaceted engagement for urban sustainable futures in Melbourne and Southeast Asia

Colin Fudge and John Fien

> The academy must become a more vigorous partner in the search for answers to our most pressing social, civic, economic and moral problems, and must reaffirm its historic commitment to what I call the scholarship of engagement.
>
> Boyer (1996: 11)

Overview

This chapter examines the range of activities in the sphere of urban sustainable futures that Royal Melbourne Institute of Technology (RMIT) University has undertaken with partners over the last five years. It elaborates the strategic direction of RMIT and the rationale behind its urban engagement in Melbourne and in the other cities where RMIT has a presence—Ho Chi Minh City, Hanoi, Barcelona, Singapore, Hong Kong, Kuala Lumpur, Shanghai and, soon, Jakarta.

This exploration of knowledge co-production covers the university's strategic direction and transdisciplinary approach to urban futures; its own sustainability policy and practices as the largest landowner in the Melbourne CBD (city business district); its multiple engagements with internal and external stakeholders; and examples of engaged research in five fields—sustainable enterprise, Indigenous reconciliation, design and social responsibility, and climate change adaptation and disaster risk reduction. From the review of these activities several substantive issues and themes are raised in relation to university–city engagement, including: the strategic direction of the University of being "global," "connected," and "urban"; implications for the nature of research; learning from the international comparative experience; and the role of the academic in engaging with the complexity of practice and a large range of stakeholders in engaged research. There are both opportunities and limits for universities and their staff in seeking to work in these ways, especially within the philosophy of knowledge co-production with government, business, and civil society.

The urban setting

Sustainability is the most significant challenge facing urban policy-makers in the twenty-first century: How can the increasing needs of the increasing numbers of urban dwellers be met in ways that do not overshoot the limits of the natural resources upon which economic and social development depends? This is a particularly important challenge across the growing Asian economies, where rapid urbanization is a key driver for economic growth and poverty alleviation. In 1950, around 10 percent of people in Asia lived in urban areas. In 2010, it was approaching 40 percent and it is expected to exceed 50 percent by 2025. These cities are also rapidly increasing in size and number. At the beginning of the last century, there were only 11 megacities, cities with populations of more than one million people, in the world. By 2030, there will be more than 500 megacities, and more than half of these will be in Asia. In addition, with the peri-urban areas of these cities rapidly expanding, Asia is now home to almost half of all urbanites on Earth and its urban population is more than three times that of Europe, the world region with the second largest urban population (Asian Development Bank 2012a: xxxvii).

Thus, effective governance, policy, and action programs for sustainable urban futures are key needs across Asia. Yet applied urban research—and the capacity to undertake it—are critically lacking. Such research provides a vital evidence base for policies and strategies to address the environmental and quality-of-life problems brought by rapid, unplanned urban growth. In particular, research is needed on ways of sustaining urban areas as engines of growth, bridging supply and demand gaps in infrastructure services, and strengthening urban management capacity (Asian Development Bank 2012b).

Within the Asia-Pacific region, Australia is notable as a highly urbanized country, experiencing rapid urban growth while still supporting prosperous lifestyles. Eighty percent of Australians live in cities and contribute over 80 percent of the GDP and employment growth. By 2029, 3.2 million additional homes will be required and by 2040 the number of urban Australians will double. However, a recent national report, *State of Australian Cities 2012*, indicates that the country's urban-based prosperity is being undermined by the side-effects of success, including population growth, social unease, road congestion, resource scarcity, and escalating housing costs (Department of Infrastructure and Transport 2012). Yet, conventional responses, such as expanding urban boundaries and existing infrastructure, would only make Australian cities increasingly dysfunctional, congested, and unaffordable. Indeed, these responses are the cause of many current urban problems. Thus, modeling reveals that maintaining the urban status quo would cost the Australian economy 480 trillion dollars over the next 50 years. Without better city planning, progressive urbanization will become a major constraint to economic, environmental, and social prosperity. It is surprising that some of the policy and practice developments in Europe (Fudge *et al.* 1996; Fudge and Rowe 2000) that address these issues have not been taken up in Australia (Horne and Fudge 2014).

A university for the urban age

This chapter focuses on the urban opportunities and challenges in Southeast Asia and Australia from the perspective of a university committed to urban education and research. Royal Melbourne Institute of Technology (RMIT) University is a globally focused university of technology and design that derives its identity and mission from its deep roots in both its urban settings and its Asia-Pacific location.

The university was founded as a metropolitan college, providing education in science, technology, and the arts for working people in the nineteenth-century Australian colony of Victoria. Donations from members of the public provided more than half the original building costs, with trade unions contributing a substantial portion. From these beginnings, RMIT has linked its development and focus to the dynamic life and growth of Melbourne, now ranked very highly among the world's most livable cities. This link is reflected in close relationships with civic, business, and community leaders in the city, as well as in RMIT's academic profile and its focus on work-integrated learning and education for the professions. RMIT is now the largest non-government holder of space in central Melbourne CBD, occupying nearly seven percent of CBD office space in what has become known as the "RMIT Quarter" of the city.

RMIT also has a presence in Asia, with two campuses in Vietnam, a permanent presence in Tianjin, major new campus-scale initiatives in Singapore and Jakarta, and a new European hub in Barcelona. The university also has numerous transnational education partners delivering RMIT courses and research in China, India, Japan, Laos, Malaysia, Sri Lanka, and Singapore in cities such as Colombo, Hong Kong, Kuala Lumpur, Shanghai and Tokyo. Altogether, RMIT has nearly 32,000 international students out of a total of over 82,000 students; for instance 12 percent of all RMIT students have a Singapore home address. As a result, RMIT academics have developed strong teaching and research linkages with industry, municipalities, international agencies, and community-based organizations in almost all Asian countries as well as many of the small island states of the Pacific.

In all these relationships, RMIT's teaching and research seek to address the global challenges of the urban age by working in partnership with others. This reflects the strategic plan, *Transforming the Future*, in which the university commits itself to orient all teaching, research, and engagement activities to the three goals of being:

- *Global* in attitude, action and presence, offering our students a global passport to learning and work.
- *Urban* in orientation and creativity, shaping sustainable cities and drawing inspiration from the challenges and opportunities they provide.
- *Connected* through active partnerships with professions, industries and organizations to support the quality, reach and impact of our education and research (www.rmit.edu.au/about/strategy).

The social relevance and service orientation of these goals are a reflection of two significant trends in higher education over the past two decades which have had profound effects on the way RMIT University operates. The first is the influence of Ernest Boyer's thinking about the "scholarship of engagement." In a 1990 report for the Carnegie Foundation, entitled *Scholarship Reconsidered: Priorities of the Professoriate*, Boyer argued that scholarship should be understood as civic engagement (Boyer 1990). He outlined what this meant in a later paper in which, like Russell Jacoby in *The Last Intellectuals: American Culture in the Age of Academe*, he lamented the way in which "being an intellectual has come to mean being in the university and holding a faculty appointment (preferably a tenured one), writing in a certain style understood only by one's peers, and conforming to an academic rewards system that encourages dis-engagement" from society (Boyer 1996: 22).

In response, Boyer argued for a "scholarship of engagement" that integrates four "interlocking functions" of academic staff. The first, the *scholarship of discovery*, involves basic research—pushing back the frontiers of knowledge through theoretical and methodological contributions to the disciplines. The *scholarship of integration* involves placing such discoveries within a larger context through applied interdisciplinary investigations. The *scholarship of sharing/teaching* recognizes the educational role of the academia not only for enrolled students but also for other audiences through community education and professional development programs. Finally, Boyer advocated the *application of knowledge* as a reflective practice in which theory and practice inform each other in the personal and professional lives of scholars. In this chapter, researcher-initiated co-production will be framed in terms of different types of scholarly engagement, including educational, interdisciplinary and transdisciplinary examples and approaches to co-production.

The second trend is the rapidly growing emphasis on sustainability in higher education. This includes a commitment to the efficient use of natural resources in campus operations and the management of campus estates, and the integration of sustainability into teaching and research programs (Cortese 2003; HEA 2012). In many cases, the sustainability ethos has also refocused and strengthened university engagement with partners in civil society, the private sector, and governments in the pursuit of enhanced social equity, productivity, and conservation. Such engagement has integrated the service learning, course guidance, and commercial relationships that have underpinned conventional university–community relations into multidimensional collaborations aimed at meeting the challenges of the sustainability transition. The research, teaching, integration, and application aspects of such sustainability-focused activities intersect with the four dimensions of Boyer's scholarship of engagement and also give practical expression to what could be considered normative aspirations.

Examples of the four dimensions of the scholarship of urban engagement at RMIT are illustrated in Figure 6.1. These include the establishment of two cross-university research Institutes, the Global Cities and Design Research

The Scholarship of Discovery

Formation of a Global Cities Institute with research to develop the conceptual foundations, theory, and practice of:

- Human security
- Disaster resilience
- Community safety
- Urban and social reconstruction post-conflict
- Green building technologies
- Climate change adaptation
- Low-carbon regional transitions
- Skills needs in industrial change
- Learning cities and regions
- Life-time affordable housing
- Sustainable design systems for housing programs for marginalized communities
- Community sustainability
- Sharing economies
- Environmental citizenship
- Urban media ethnography
- Social practice and conservation
- Indigeneity, sport and health

The Scholarship of Integration

PhD research in central coastal Vietnam on issues such as:

- Design and construction of culturally-relevant and typhoon resilient housing in low-income neighbourhoods;
- Design of sustainable medium density housing models to replace deteriorating Soviet-era apartment blocks; and
- Development of a multi-objective decision support system for climate change adaptation planning.

Development of significant on-line tools for international agencies, e.g.:

- UNESCO – Teaching and Learning for a Sustainable Future
- UNDP – Development Communication Toolbox
- Rockefeller Foundation – Better Evaluation Toolbox

Development of "Circle of Sustainability" methodology for the UN Global Compact for Cities program replacing rigid metrics approach to sustainability indicators

The Scholarship of Sharing through Engaged Teaching

Comprehensive Work-Integrated Learning in all degree programs in Melbourne and Vietnam.

Social Enterprise studies available as a cross-university "minor" in all degree programs.

Development of City Design Labs in partner cities, which integrate students into city agencies to develop planning and design solutions for issues and problems identified by the agencies.

Cross-disciplinary "Building the Community" integrated practice studio in Hoi An, Vietnam.

Development of PhD Summer School program on urban sustainability for the Asia Pacific Network for Postgraduate Sustainability Education and Research (ProSPER.Net): Ho Chi Minh City (2010), Tokyo (2011), Bandung (2012).

Provision of training courses on cross-cultural work practices and environments for transnational companies operating in SE Asia.

The Scholarship of Application through Reflection

Formation of a Design Research Institute that has critically reflective practice-based research as its *modus operandi*.

Invitational Masters and PhD programs in Architecture based upon the development of reflective practice by and amongst leading professionals.

Development of a Design Archive to preserve plans of significant Melbourne buildings for education and research purposes.

Cross-degree core course on sustainability in the School of Global, Urban and Social Studies requires students to research and propose solutions to on-campus sustainability problems.

Integration of minimum standards of eco-design for energy efficiency in all campus renovations and new buildings.

Rigorous reporting program regarding performance on sustainability indicators agreed through multiple declarations and agreements, e.g. The Australian Technology Network aim of reducing carbon emissions by 25% by 2020.

Figure 6.1 Sustainable urban futures examples of Boyer's four dimensions in a "Scholarship of Engagement" at RMIT University

Institutes; the integration of learning, research, and engagement projects in City Design Labs; the construction of a purpose-built Design Hub to locate all design researchers in a shared space; the establishment of partnership projects with government agencies to develop guidelines for place-making in urban renewal, health, equity, and noxious industry; a National Covenant for Sustainable Packaging; and "green loans" to support the sustainable renovation of residential properties. Internationally, it involves RMIT's Global Cities Institute hosting the United Nations Global Compact for Cities process, being a partner university with UN-HABITAT, and research on topics as diverse as skills for a green economy for the Asian Development Bank and teaching climate change for UNESCO in Indonesia.

However, a key lesson learned from the scholarship of engagement at RMIT, as elsewhere, is that working within the different dimensions over several years integrates them to form an emergent type of university and academic. Indeed, as Spanier argues, it is only through the synergies that come from "integrating the teaching, research, and public service missions of universities that … we will create and support the broad-based and active learning community that is best prepared to cope with society's challenges" (1997: 8). The nature of learning and research in this emerging type of university is analyzed in the final section of this chapter. However, first, the chapter outlines five examples of the multidimensional engagement of RMIT in Melbourne and Southeast Asia. These are: (a) the integration of teaching and research into a comprehensive retrofit of the entire university estate through a Green Government Buildings Project; (b) a social entrepreneurship program; (c) an Indigenous research support program known as the "Koori

Cohort"; (d) a focus on design, urbanism and social responsibility through "live projects"; and (e) embedding urban research in policy and planning reform.

Green Government Buildings Project

The integration of the four dimensions of a scholarship of engagement is evident in the Green Government Buildings Project. RMIT University is undertaking an eight-year sustainability transformation process to retrofit all buildings for maximum energy and water performance and reduction of carbon emissions. This is being done not as an engineering project *per se* but to find ways of enhancing the university's core business of teaching, research, and community outreach through transformed environmental performance and outcomes. The strategy was developed in response to a provincial government Green Government Building Program (GGBP) that invited applicants to submit proposals for transforming government buildings (such as hospitals, schools, universities, office blocks, etc.) into sustainability "demonstration sites" of that category of facility. Successful applicants were offered an interest-free government loan to the value of the projected utility savings over an eight-year pay-back period. RMIT University invited tenders from environmental services firms to develop strategies for retrofitting all its built assets but stipulated that teaching, research, and outreach were to be central aspects of submissions. Two engineering firms, Siemens and Honeywell, were selected to progress the proposal to the next stage of development.

A one-year period of university-wide consultation and planning followed. This revealed great enthusiasm from staff across academic, property services/facilities management, information technology, and administrative sections of the university, and plans were made for a wide range of re-engineering, educational, and research projects. However, while the total utility savings over the proposed eight-year period would be AUD29 million, the total cost of the retrofitting would exceed AUD100 million. Thus, in addition to an interest-free government loan for the AUD29 million, to be repaid through utility savings under the GGBP, RMIT University has borrowed the balance to fully undertake what may be one of the largest precinct-scale retrofit programs ever.

The scope of works on the central city campus in Melbourne includes:

- A Green Precinct Program to provide critical infrastructure upgrades across the campus to transition RMIT to a precinct-based concept.
- A Green Energy Program to provide clean energy for the university.
- A Green Air program to deliver clean, well-controlled, healthy air-conditioning to all campus buildings.
- A Green Light program to provide a comprehensive approach to lighting that is smarter, more efficient, and requires much less maintenance while reducing the number of fittings used.

- A Green Water Program that applies best-practice water management principles to holistically address water supply and demand.

Funded *within* this work is a complementary, but integral, teaching and research program based upon:

- A Green Learning and Teaching Program to create sustainability curricula and teaching resources aimed at engaging students across disciplines in deeper action-based learning experiences.
- A Green Research Program that will transform the City Campus into a living laboratory to harness knowledge created by the project and advance research into sustainable urban futures.
- A Green Engagement Program to promote interactive dialogue with students and staff to foster a sustainability-inspired culture.
- A Green Careers Program to incorporate graduate recruitment pathways, work integrated learning, and opportunities for RMIT students to become work ready via leadership and mentoring opportunities.
- A Green Innovations Program to enhance education, engage the community, and promote a culture of sustainability.

Figure 6.2 illustrates the range of activities in these teaching and research programs at the Central city campus. These works and innovations in teaching and research will cost approximately AUD50 million with a similar scope of works and activities being undertaken on other RMIT campuses.

RMIT Social Entrepreneurship Program

The integration of the scholarships of discovery, integration, sharing/teaching, and application is reflected in the development of a wide range of activities promoting social entrepreneurship and social enterprise amongst students at RMIT. In the early 2000s, RMIT University appointed Australia's first Professors of both Youth Studies and Sustainability. One of them built teams of researchers in twelve countries to investigate, compare, and contrast youth environmentalism in the Asia-Pacific region and identify the needed educational responses to this (Fien, Yencken and Sykes 2002). Both professors jointly developed a community development process for enhancing youth leadership for sustainable consumption at local and regional levels (Fien, Neil and Bentley 2008). Thus, faced with the tensions that young adults experience between idealism on social and environmental matters and concerns about future employment, it is not surprising that the university has responded by promoting student interest in and knowledge about social entrepreneurship.

This is done through the Social Enterprise Group at RMIT, which supports a range of business development, research, and educational initiatives designed to foster solutions to social and environmental problems by promoting

Green Program	Project	Description
GREEN LEARNING & TEACHING	LP1.1	From Concept to Action: Designing a Full Integrated Sustainable Minor in Social Entrepreneurship
	LP1.2	The Matter of Landscape: Sustainable Design Strategies for RMIT City Campus
	LP1.3	The Green Office and Clasroom: Designing Social Interventions from Analysis of Green Living Monitoring at RMIT University
	LP1.4	Playful Triggers: Raising Student and Staff Sustainability Awareness and Action
	LP1.5	Green Print
	LP1.6	An Interactive Teaching Laboratory for Sustainable Electrical Building Service Technologies and Systems
	LP2.1	The Greenhouse Challenge: Promoting Sustainability in Work-Integrated Learning
	LP2.2	Showcasing IT for Sustainability: RMIT Case Series
	LP2.3	Navigating a 'Fierce Planet': Augmenting Sustainability Learning through Gaming and Simulation
GREEN RESEARCH	RP1.1	Buildings Engineered for Urban Sustainability: Performance Assessment, Forecasting, and a Toolkit for Selection of Performance-Based Retrofits
	RP1.2	Inhabiting Buildings: Embedding Sustainability into RMIT Culture
	RP1.3	Powering Our Future Cities: Achieving Greener Government Buildings and Energy Use Behaviour Using Demand Response and Energy Information
	RP1.4	Enhanced Micro-Grid with Sustainable Energy Storage for Peak Load Levelling
	RP1.5	Online Infrastructure for iCO2mmunity: Personal and Community Carbon Footprint Monitoring for University-Wide Engagement towards Greener Living
GREEN ENGAGEMENT	EP1.1	Lucid Dashboard (in collaboration with Honeywell)
	EP1.2	RMIT Sustainable Futures Academy
	EP1.3	GGB Content & Marketing Strategy Support
	EP1.4	RMIT Teaching Awards: 'Excellence in Sustainability' Awards Categories
	EP1.5	Siemens Guest Lecturer for RMIT at Sustainable Living Festival
	EP1.6	Siemens Assistance for RMIT External Sustainability Awards Applications
GREEN CAREERS	CP1.1	RMIT Siemens Graduate Recruitment & Engagement Strategy
	CP1.2	RMIT 'LEAD' Program: GGB Sustainability Showcase Initiative
	CP1.3	RMIT Siemens Student Mentor Program
	CP1.4	Siemens Work-Integrated Learning Program
GREEN INNOVATIONS	IP1.1	Electric Racecar Powered by Renewable Energy
	IP1.2	Piezo Electric Demonstration
	IP1.3	Electric Vehicle Charge Stations

Figure 6.2 Education and research programs in the Green Government Building Project on the RMIT city campus

innovation and entrepreneurship among students, staff, and alumni. The activities of this group include:

- Hosting an Entrepreneur in Residence program, the only one of its type in Australia. The program offers students access to an experienced business person, consultant, and entrepreneur who can provide individual advice about their concepts for new or existing businesses. The Entrepreneur in Residence also contributes to formal lecture programs as well as less formal activities such as alumni workshops and an Entrepreneurs' Coffee Club, a networking process formed to encourage discussion about new business ideas, start-ups, and running successful ventures.
- Organizing a Business Plan Competition open to all students studying at RMIT in Australia and at offshore campuses, which encourages student teams to create innovative new ideas and businesses by developing competitive business plans. To help students develop the necessary skills, the Business Plan Competition runs a series of free workshops—led by professionals running their own companies—in the areas of Business Planning, Marketing, and Financial Planning.
- Managing RMIT SEEDS, a grants and fellowship program that supports the development of student-run social enterprises, both locally and internationally. The focus is on RMIT student-led projects that are innovative, sustainable, and have a positive impact upon the social and environmental needs of individuals and communities. The program has provided start-up funds for ventures such as: *WATT ELSE*, in which an Engineering and Sustainable Energy master's student developed a business venture that provides cheap solar panels to communities; *otherWISE*, in which two Ph.D. students developed a company for youth networking and training for sustainability; and an *African Women's Sewing Enterprise*, which was developed by a partnership between a textile design student and Somalian refugee to enable a group of African women to recycle and up-cycle old textiles that are then sold through a Green Collective.

Sponsored by the RMIT Sustainability committee, and developed through the Green Government Building Project, the Social Enterprise Group has also developed a four-course "minor" suite of studies which will be available to all students across the university. The minor will use case studies of, and work experience in, social enterprise firms to help students integrate small business planning and management skills into their professional studies (for example, within architecture, fashion, education, law, or health sciences) within the framework of sustainability and social justice. Two courses in the minor will require students to work in small teams to plan, implement and evaluate a social enterprise they have created. Business leaders, who are members of the Industry Advisory Committee to the School of Management, which sponsors the Social Enterprise Group, have recommended that the "minor" be made a core component of the School's long-standing Bachelor of Business

(Entrepreneurship) degree and that it be renamed as the Bachelor of Business (Innovation, Entrepreneurship, and Sustainability) degree.

The Koori Cohort of Researchers

Reconciliation with Australia's Indigenous people forms a key part of RMIT's social platform. Research programs have supported the growth of Indigenous cultural festivals as sustainable enterprises, mentored Indigenous youth through community resilience projects, promoted the market for Indigenous art, and helped develop consultation, design, and construction protocols for sustainable housing systems in remote Indigenous communities. Operating under a credo of "no survey without service," these researchers have also provided *pro bono* services, including preparing housing condition reports for an Indigenous town after a cyclone, preparing design and construction portfolios to facilitate fundraising for three remote Indigenous cultural centers, and facilitating design charettes to help a community respond to government proposals for a remodeled town center. In addition, in partnership with the Plumbers Union, RMIT has sponsored the training of a cohort of Indigenous plumbing apprentices while the RMIT Community Services team is providing vocational qualifications in community health for Indigenous people in their communities.

A significant program for Indigenous students is the "Koori Cohort of Researchers," which helps Indigenous community members undertake research degrees at the master's and doctoral levels (Brearley 2010). Examples of research questions that have been explored or are being investigated include:

- In what ways does my artwork reveal my identity as an urban Indigenous artist and raise awareness about contemporary Indigenous issues?
- How do young Koories build, maintain, and strengthen their cultural identity?
- What factors have influenced the establishment and continuity of the Woolum Bellum Koori school?
- What are the principal challenges and pressures facing University Indigenous Unit Managers?
- How can a Master of Business Administration degree be aligned with the capacity building needs of the Aboriginal and Torres Strait Islander community?

These examples illustrate that members of the Koori Cohort, which includes artists, musicians, educators, media specialists, and community development practitioners, conduct their research through Indigenous methodologies and within Indigenous ways of knowing and knowledge systems. Paramount among these is the Deep Listening process, as both a research method and a framework for presenting the outcomes of research projects to the community. Thus, the Creative Events Program within the Deep Listening Project has arranged a series of exhibitions and performances that combine art, music, multi-media, and Indigenous culture by Indigenous and non-Indigenous musicians, performers, and artists who are also researchers. The aim of these events is not only to celebrate Indigenous culture but also to revitalize language through the Arts, and to re-present Indigenous culture and ways of knowing to wider Indigenous and non-Indigenous communities.

Building the community: design and social responsibility

The emphasis on urban social sustainability in these Koori Cohort and Social Enterprise programs is also characteristic of a range of design research and teaching projects at RMIT University. For example, nearly twenty years ago, staff in the landscape architecture program were taking students into Indigenous communities to work in collaborative ways to help produce "landscapes for health" through improvements in settlement planning (Sinatra and Murphy 1997). This work has continued not only in Indigenous communities, as outlined above, but also in Melbourne and Southeast Asia. Indeed, building on the design–build traditions of the Rural Studio (Dean and Hursley 2002) and the Sheffield School of Architecture (Sara 2006), RMIT staff have conducted numerous design studios as "live projects" and as vehicles for engaged design research.

As Dodds, Harrisson, and Charlesworth (2012) write, live projects engage with real communities and reflect the growing social and ethical commitment to expand the role of design education beyond the university. Some of these projects have a commercial base, albeit on a low budget, through a "Project Office" in which staff and students consult to industry, community groups, and government. Others address the design needs of marginalized communities. Beginning as a research and teaching cluster under the acronym "CODE" (Community-Oriented Design Engagement), a large number of studios have addressed design solutions to:

- dilemmas arising from life in a poorly designed and located high-rise tower for social housing in the inner Melbourne suburb of Fitzroy, which had developed into a dysfunctional complex with significant safety issues chiefly related to drug dealing and use;
- the need for community-controlled health facilities in Indigenous communities;
- neighborhood renewal and social capital outcomes for a low-income suburb on the outskirts of Melbourne;
- problems of food insecurity in low-income neighborhoods (addressed through community gardens); and
- low levels of provision and use of public transport in Melbourne.

These and other projects have been the focus of documentation and reflection, leading to the publication of *Live Projects: Designing with People* (Dodd, Harrisson, and Charlesworth 2012), which analyses the rationale, learning processes, outcomes and, ultimately, the effectiveness of live projects within the university setting.

Staff in the environmental science and planning programs at RMIT University have made the Vietnam cities of Hanoi and Ho Chi Minh City the focus of applied problem-solving projects in which students have engaged with issues such as climate change, waste management, water pollution, and traffic congestion, and have presented policy and program solutions to various national, provincial, and local governments. Among the more significant of student projects in Vietnam has been a series of design studios in the city of Hoi An called "Building the Community." Over a period of five years, the deep immersion of RMIT staff from architecture, construction, sustainability, and communication disciplines in the needs of community organizations and the People's Committee (local government), has resulted in community-based designs for transitional housing for homeless youth, a disability day-care center, and flood and climate-change adaptation plans for the city. The aim of each of these projects in Hoi An was the eventual construction of the students' designs. Therefore, a key component of "Building the Community" was for students not only to consult with potential users (of the buildings or plans being developed) but also with local government officials, to understand the politics of decision making in Vietnam, and with contractors and quantity surveyors, to establish an accurate cost estimate for the designed works. Thus, in each case, students produced a costed scoping document that could be presented to potential sponsors of the project. The students in one of the "Building the Community" projects also designed a web-based manual to help students and lecturers at other universities develop similar projects of their own.

The first of these projects to be actually built, the Dien Bahn Disability Day-Care Center, opened in 2012, following financial support from RMIT for its construction. The Center caters for eighty children who were born with disabilities attributed to the large concentration of Agent Orange herbicide in the soil in this area, a legacy of the American War in this region. These children, aged 2–18, attend the Center on a daily basis to receive physiotherapy and take part in special education programs provided by an English NGO. This has the potential to transform their future education and employment possibilities, as well as that of their wider family groups. It is envisaged that RMIT students from Melbourne and Vietnam will undertake Work Integrated Learning projects at the Center, providing services such as office management, IT infrastructure, nursing, and education according to their course specializations (RMIT Architecture 2012).

Reflecting on the scholarship of engagement that informs these design studios, Dodd (2012) draws upon the literature of reflective practice to note three benefits of teaching and learning through live projects. First, students are able to learn *through* rather than *for* professional action, and do so within the relative safety of an educational context. Second, they are able to engage with the critical moral issues of the contemporary world—social marginalization, dysfunctional urban environments, climate change, etc. Third, live projects are a way for design educationalists to connect the need to "'rehearse' design as an iterative process in an academic setting, but with reference to the real dimensions of practice and application" (p. 22). In this regard, she draws upon Lave and Wenger's work on situated learning to argue that such reflection focuses less on the conceptual and cognitive processes and outcomes of learning, and more on the "kinds of social engagements [that] provide the proper context for learning to take place" (Lave and Wenger, 1991: 14). Harrisson (2012) describes the transformational potential of this for both lecturers and students (p. 159). In this way, live projects can be seen as a key example of RMIT's commitment to facilitating sustainable urban futures for vulnerable communities in Melbourne and Vietnam.

Urban research for policy and planning

The transformational orientation of the teaching and research projects outlined in the previous section is a feature of the way much urban research at RMIT exemplifies a systematic approach to knowledge co-production by embedding activities and engaging research in local policy and planning. Research within the Sustainable Urban Futures Research Program at RMIT (SURF–RMIT) is driven by the question: How can people improve human well-being in a world of increasing urban concentration, given the economic, environmental, social, and political challenges facing future development? This section outlines the policy and planning outcomes of examples of research within the SURF–RMIT program. These include studies of affordable housing, climate change adaptation, and disaster risk management.

Research on affordable housing

One RMIT research team was commissioned to produce the housing component of the Victorian Government report, *A Fairer Victoria*, which was an action plan to drive investment and economic development over a one-year period for the most disadvantaged people and places in Victoria. Having analyzed trends in housing affordability in Victoria, the team identified growing housing affordability stress among low-income Victorian households and increasing social and economic polarization. They participated in the workshops of the Policy and Strategy Projects Group of the Department of Premier and Cabinet, and made three formal presentations. The outcomes of the research and workshops helped prompt the Victorian Government to invest AUD49.6 million over five years in measures to increase the availability of affordable housing in Victoria. Two members of the team were also appointed to the Advisory Committee to Victorian Office of Housing to develop and implement a set of housing affordability indicators to monitor policy impacts against the strategy established in *A Fairer Victoria*.

Since then, the team has expanded in membership and focus, and developed the concept of "lifetime affordable housing." Until recently, policy debate around equity and energy efficiency was frequently simplified to an argument of "sustainability vs. affordability." Research on "lifetime affordable housing" developed a methodology for considering the full lifecycle costs of housing provision, including not only upfront capital costs but also ongoing energy and transport costs. The analysis of the cost implications of higher thermal efficiency, locational efficiency factors and the social cost implications of these across socio-economic groups contributed evidence that has informed policy responses and new programs in Victorian social housing provision and design guidelines for new housing estates. The lifetime affordable housing methodology has also been applied in an evaluation of house designs for remote Indigenous communities in Australia. Additionally, it is being incorporated into the design and construction of medium-density low-income housing in Vietnam through a capacity-building partnership with the Vietnam Green Building Council and with architecture staff at the University of Da Nang.

Research on climate change adaptation and disaster risk reduction

In their various ways, Southeast Australia, where Melbourne is located, and coastal Vietnam are among the most vulnerable regions in the world to the impacts of climate change. The increased typhoon and flood risks in Vietnam over the past two decades and predicted temperature and sea-level rises are well-documented (IPONRE 2009; UNISDR 2011). These call for engaged research and urban adaptation programs. Less well known is the situation of Melbourne, whose peri-urban areas are already among the most vulnerable in the world to bushfires. Climate change is bringing increases in temperatures, decreases in rainfall, prolonged drought, and further heightened bushfire risk (DSE 2008). This risk is intensified by intensive suburbanization, which is causing Melbourne's peri-urban areas to increase in population at twice the rate of the rest of regional Victoria (Buxton, Haynes, Lechner, and Butt 2009). Engaged research on climate change adaptation and disaster risk reduction is needed in Melbourne and Vietnam, as well as elsewhere, to provide the land use planning policies and guidelines needed to reduce vulnerability and to provide the tools that local governments and communities need to increase their adaptive capacities. Three examples of RMIT projects that are seeking to respond to these needs are: studies of community safety and bushfire risk in Melbourne, the development of local government and coastal adaptation planning tools, and adaptation planning in coastal Vietnam.

A decade of research on bushfires and community safety at RMIT (Handmer and Haynes 2008) was called upon following devastating firestorms to the east and north of Melbourne on February 7, 2009, later named "Black Saturday". The fires burned 450,000 hectares, destroying more than 3,500 buildings. In addition, 173 people were killed, 414 more were injured, and a further 7,562 were displaced, and more than a million animals perished. In the preceding month an additional 374 people had died from heat stress (Fisher 2012). The Victorian Bushfire Royal Commission that was established to investigate the fires engaged the RMIT Centre for Risk and Community

Safety to conduct intensive interview research in the weeks after the fires and also to review all related research (Whittaker and Handmer 2010). Their findings indicated limited community understanding of warning codes and systems, and that the high level of claimed awareness of bushfire risk among those living in vulnerable areas was not matched by an understanding that *their* homes and families were at risk. Hence, while 60 percent of households claimed to have fire plans, less than a quarter of these plans were actually written down and even fewer had been rehearsed. This finding challenged the existing policy frameworks that encouraged families to make plans to "leave early or stay and defend" and has led to reconsiderations of shared responsibilities in emergency preparation and renewed community education and warning programs.

Further RMIT research on the "Black Saturday" fires indicted "a clear demonstration of … 'regulatory failure' under the current Planning System" (Buxton *et al.* 2009: 32) as Victoria had no statutory requirements for risk management for climate change in land use planning or approval processes. As a result, the team argued for the adoption of an anticipatory approach to risk management within the planning system based upon the precautionary principle and for a review of the capacity of the relevant controls and authorities to manage such risk through state-wide planning reforms.

To assist in this, another RMIT planning team is developing tools to assist local authorities with integrating climate change adaptation into their planning processes. One of these, the Climate Change Adaptation Toolkit, was developed by RMIT University together with NetBalance and the City of Greater Geelong and provides a three-step process that helps organizations integrate adaptation into decision-making under the conditions of complexity and uncertainty posed by climate change (see Figure 6.3). Another tool, the "Climate Change Adaptation Navigator" (www.adaptation-navigator.org.au) visualizes the adaptation process as a journey during which information and skills pertaining to different aspects of climate change adaptation processes can be discovered, including:

- Ten different stages of an adaptation planning process, represented as imaginary geographic areas on a mental map;
- Climate science and what organizations need to adapt to; and
- How to make decisions for adaptation and how to "do" adaptation.

Although too new to be evaluated at the time of writing, the "Navigator" contains case studies of the adaptation journeys taken by the local governments that participated in its development.

Several climate-change-adaptation and disaster-risk-reduction projects are also underway in Asia and the Pacific. These include studies of integrated community, housing, and infrastructure projects in Sri Lanka, India, and Aceh following the 2004 Indian Ocean Tsunami; the development of tools for evaluating post-disaster housing projects in the Pacific and strategies for

Figure 6.3 Climate change adaptation toolkit.
Source: Available on-line at http://www.netbalance.com/s/Toolkit.pdf.

enhancing climate adaptation policy and planning in Vietnam; and the development of adaptive capacity for typhoons and flooding in areas ranging from Dhaka slums to the Mekong Delta and central coastal Vietnam. The latter studies include doctoral research at RMIT by senior Vietnamese officials on the design of low-cost housing in flood-prone areas along the Hue River; strategies for vulnerability assessment and modeling of social, economic, and infrastructural adaptation to flooding around Da Nang; and the development of women- and family-focused strategies for reducing the impacts of flooding in Ho Chi Minh City. These partnerships between RMIT Melbourne and Vietnamese researchers help to disseminate and build capacity in the contextualization and use of adaptation and disaster-risk-reduction tools from Australia to Southeast Asia.

Conclusion

The engagement of RMIT University with its urban settings reflects the integrated nature of the scholarship of engagement. While the examples listed in Figure 6.1 illustrate the scholarships of discovery, integration, sharing, and application, the five cases studies of university–city engagement show how these concrete examples of co-production are best seen as an integrated and multifaceted activity. However, reflection on the case studies reveals a number of substantive issues and themes about university–city engagement. For example, despite the breadth of the examples, and their presence across many colleges and schools, a key issue is the need to up-scale engagement practices across the university and to shift innovative practice into the mainstream of university activities.

This is a capacity-building challenge not just for staff on an individual level, but also in terms of the support structures and enablers that the university provides to encourage the development of a university-wide scholarship of engagement. Such enablers could include allocating resources to enable researchers to travel more frequently between Australia and research sites in Southeast Asia and a swifter roll-out of e-learning infrastructure to free academic staff from high class-contact hours and the need for weekly face-to-face teaching. It could also include a reconsideration of the criteria for promotion from the traditional concepts of "teaching, research, and service" to "learning, innovation, and engagement" (adapted from Kellogg 1999), to more closely parallel the spirit of the scholarship of engagement.

Despite such concerns, a number of important lessons about the opportunities and limits of university–city engagement emerge from the case studies. These concern (a) defining the field of urban studies, (b) the practice of engaged, interdisciplinary research, (c) processes for engaged research within Asian cities, and (d) the need to develop formal partnerships between cities and universities as "urban knowledge arenas."

First, there is a need to consolidate the field of urban studies as a transdisciplinary endeavor. Knowledge about the complexities of cities, the making of urbanscapes as district and precinct entities, and urban residents and the services they perform and need is not the domain of any one discipline (e.g. architecture, planning, geography, or sociology). Moreover, as "the urban" is not central to any of these fields of study, urban studies—in the words of Nolmark—"too often falls into the cracks between the disciplinary fault lines" (2009: 18). Nolmark goes on to argue that:

> It should thus be no surprise that the desire for disciplinary integration has proved rather difficult in practice; academia becomes ever more specialized and thus lacks the traditions for synthesis of various aspects of

> the research required to develop a more rounded understanding of the urban context … Unfortunately, there is no single "science" of urban studies that is able to encompass all the different aspects of the urban situation.
>
> (Ibid.)

The answer to this dilemma may reside in a much greater recognition of transdisciplinarity as the epistemological base of urban studies and the creation, perhaps, of an emergent knowledge field. Such an approach could be the solution to the disciplinary chauvinisms and logistical difficulties involved in successfully discovering, integrating, applying, and sharing knowledge in interdisciplinary contexts. It is too early to trace the nature of transdisciplinary urban studies, but among its characteristics would be viewing praxis as the basis of engaged scholarship to address potential theory–practice gaps; ensuring that praxis is guided by the moral imperatives of sustainability—equity, conservation, democracy, and appropriate development; maintaining a futures orientation; promoting a dialogical, mutually supportive relationship between universities and their urban settings; and allowing the boundaries to be blurred between knowledge creators (the university) and knowledge users (the city) in knowledge co-production.

These characteristics point to changes in the nature of research and the role of the academic in response to the complexity of practice-embedded research and the need to manage a very wide range of stakeholders in knowledge co-production. Thus, the practice of engaged, interdisciplinary urban research "points to the need to conceive research and knowledge as related to and involved in society, not prior to or above social life and its realities" (Nolmark 2009: 17). In contrast with research that stands outside the urban management process and seeks to identify failings in urban systems, transdisciplinary urban research seeks to identify what does and doesn't work and why and, by investing city officials with the role of co-researcher, aims to produce guidelines and tools for enhanced urban design and management (Moback *et al.* 2012). As Polk, Kain, and Holmberg argue, "when it comes to understanding and realizing societal transformation, there is … a need to shift into modes and sites of collaboration in knowledge production and problem-solving that are closer to concrete action … [S]uch modes of working together can be understood as open system living laboratories, anchored simultaneously in academia and urban practice" (2012: 3).

This reflects the epistemological priority of "knowing how" over "knowing that" (Ryle 1949), or "Mode 2" knowledge over "Mode 1" knowledge (Gibbons *et al.* 1994). It also facilitates a more inclusive view of forms of knowledge and knowledge production in urban studies, one not restricted to the empirical. Such forms include Indigenous and feminist knowledges, arts- and design-based knowledge, and knowledge through professional practice and reflection (Duxbury, Grierson, and Waite 2007; Gray and Schubert 2010; van Schaik 2005). Thus, one RMIT project in Southeast Asia involved the development of a handbook of arts, design, and local-knowledge-based strategies for promoting sustainable livelihoods (Brennan, forthcoming).

Thirdly, the wide experience of RMIT urban researchers in many Asian cities has yielded valuable lessons on procedures for developing effective partnerships for engaged research in Asia. Chief among these is the spirit of humility, intercultural respect, and genuine interest with which research issues and partners are approached. These traits are reflected in behaviors such as a willingness to engage in active listening, to understand issues and problems from local perspectives, to put aside personal research interests and perceived solutions, and to adapt to local ways of conducting discussions, negotiations, and research. This also involves understanding and being patient with bureaucracies and relationship protocols in city and national governments, which often are best handled in partnerships with local university colleagues and business leaders. Such issues have been explored in depth by Mulligan and Natarajah (2008, 2012). However, some tension can still remain between a culturally sensitive approach and strong imperatives to provide answers to immediate issues and developing concerns.

As a result, engaged research partnerships in Asian cities take a long time to develop and require investing time and resources in developing mutual trust. Activities towards such ends may include: meeting Asian colleagues outside of work hours in their spaces, developing personal friendships, naming them as co-researchers in grant applications, providing guest lectures and training workshops, supporting doctoral scholarships for members of staff, and arranging study tours for small teams of colleagues. Because of the time and resources that such activities demand, RMIT has sought to narrow the range of cities in which it develops engaged relationships, and concentrate on building deep partnerships in the cities of Hanoi, Ho Chi Minh City, Singapore, and Hong Kong rather than all the many cities—as dispersed as Colombo, Chennai, Dili, Dhaka, and Honiara—where various staff members have been and still are working.

Finally, RMIT's experience of scholarship that engages with its city settings points to a need for more formalized knowledge partnerships than have been achieved to date. Unfortunately, Australia does not have a strong culture of evidence-based policy making, especially among local governments that have small budgets for research and often believe they have more pressing tasks than engaging too closely with universities (Banks 2009). As a result, the knowledge-city movement, based on viewing knowledge industries as drivers of economic growth (Carillo 2007), prevails over knowledge-based urban development in all Australian cities, including Melbourne. The rapid pace of urbanization in Southeast Asia, together with the capability gaps in urban policy-making and planning, renders the ideas of evidence-based policy and knowledge-based urban development virtually redundant.

Thus, there is a need to consolidate RMIT's urban engagement in Melbourne and its primary city partner sites. This may be undertaken by exploring the concept of Urban Knowledge Arenas with a wide range of urban stakeholders from varying levels of government, industry, unions, civil society, and other nearby universities to seek potential agreement on a formalizing of the relationship. According to a European Council on Science and Technology Report, an Urban Knowledge Arena "is a multidisciplinary multi-stakeholder structure with the objective of generating, managing and implementing urban knowledge" in which a collaborative team "agrees to work together in partnership to carry out a joint program" (Nolmark 2009: 36). As such, the formalized arrangement can be used not only to improve the focus and management of university–city research partnerships, but also to address the complex problems that cities face. RMIT is beginning to work towards this goal and has recently entered an agreement with the state Environmental Protection Authority (EPA) and two other universities to undertake applied research on five city problems, including health and environment, waste management, water futures, and sustainable business practices. In Vietnam, Urban Knowledge Arena type partnerships are being developed with local People's Committees, mediated through partners such as the Vietnamese Academy of Social Sciences and the Institute of Strategy and Policy on Natural Resources and Environment (Hanoi), the Southern Institute for Sustainable Development (Ho Chi Minh City), Action for the City (Hoi An), and the Quang Nam provincial Department of Natural Resources and Environment (Da Nang and Quy Nohn City). However, much still needs to be done on all these and wider urban knowledge area partnerships.

Success in achieving multifaceted engagement and co-production of knowledge between universities and their urban settings requires changing the behavior of both universities and other city stakeholders. Drawing on the experiences of engagement at the University of Salford, England, Perry and May describe these changes in terms of three important steps:

> [F]irst, the consolidation, integration and sharing of existing knowledge for the benefit of urban areas; second, better learning between places and enhanced understanding of the strategies that cities can pursue in their transitions to knowledge-based economies and societies and third, the development of more holistic, rather than piecemeal, frameworks for knowledge exchange at the urban level, which transform as well as translate and enable the up scaling of innovative practices.
>
> (2010: 21)

These steps also reflect RMIT University's pathway to a scholarship of urban engagement—and point to the nature of the work that lies ahead.

References

Asian Development Bank (2012a) *Green Urbanization in Asia Development: Key Indicators for Asia and the Pacific 2012 Special Chapter*. Manila: Asian Development Bank. Accessed August 29, 2012 at www.adb.org/sites/default/files/pub/2012/ki2012-special-chapter.pdf.

——(2012b) *Urban Development: Key Issues and trends*. at www.adb.org/themes/urban-development/issues. [Accessed August 29, 2012]

Banks, G. (2009) *Challenges of Evidence-based Policy Making*. Canberra: Commonwealth of Australia.

Boyer, E. (1990) *Scholarship Reconsidered: Priorities of the Professoriate*. New York: The Carnegie Foundation for the Advancement of Teaching.

——(1996) 'The scholarship of engagement', *Journal of Public Service and Outreach*, 1(1): 11–20.

Brearley, L. (2010) *Gulpa Ngawal: Deep Listening*. Melbourne: RMIT Press.

Brennan, L. *et al.* (eds) (forthcoming) *Growing Sustainable Communities: A Development Guide for Southeast Asia*. Melbourne: Tilde University Press.

Buxton, M., Haynes, R., Lechner, A. and Butt, A. (2009) *RMIT Submission to the Black Saturday Bushfires Royal Commission*. Accessed November 21, 2012 at http://periurban.org.au/references/Research%20Bushfires%20Royal%20Commission%20submission%20final.pdf.

Carillo, F. (ed.) (2007) *Knowledge Cities: Approaches, Experiences, and Perspectives*. New York: Butterworth–Heinemann.

Cortese, A. D. (2003) 'The critical role of higher education in creating a sustainable future', *Planning for Higher Education*, 3(3): 15–22.

Dean, A. O. and Hursley, T (2002) *Rural Studio: Samuel Mockbee and an Architecture of Decency*. Princeton, NJ: Princeton Architectural Press.

Department of Infrastructure and Transport (2012) *State of Australian Cities 2012*. Accessed December 4, 2012 at: www.infrastructure.gov.au/infrastructure/mcu/soac/index.aspx.

Department of Sustainability and Environment (DSE) (2008) *Climate Change in Victoria 2008*. Accessed November 24, 2012 at: www.climatechange.vic.gov.au/publications.

Dodd, M. (2012) 'Design, education, society: A lineage of learning by doing', in M. Dodd, F. Harrisson and E. Charlesworth (eds), *Live Projects: Designing with People*. Melbourne: RMIT Press, pp. 21–31.

Dodd, M., Harrisson, F. and Charlesworth, E. (eds) (2012) *Live Projects: Designing with People*. Melbourne: RMIT Press.

Duxbury, L., Grierson, E.M. and Waite, D. (eds) (2007) *Thinking Through Practice: Art as Research in the Academy*. Melbourne: RMIT University Press.

Fien, J., Neil, C. and Bentley, M. (2008) 'Youth can lead the way to sustainable consumption', *Journal of Education for Sustainable Development*, 2(1): 51–60.

Fien, J., Yencken, D. and Sykes, H. (eds) (2002) *Young People and the Environment: An Asia-Pacific Perspective*. Dordrecht: Kluwer.

Fisher, P. (2012) 'Melbourne', in E. Blakeley and A. Carbonell (eds), *Resilient Coastal City Regions: Planning for Climate Change in the United States and Australia*. Cambridge, MA: Lincoln Institute on Land Policy, pp. 145–80.

Fudge, C., Ludlow, D. and Pauli, S. (1996) *European Sustainable Cities*. EU Urban Environment Expert group. Brussels: European Commission.

Fudge, C. and Rowe, J (2000) *Implementing Sustainable Futures in Sweden*. Stockholm: Building Research Council on behalf of MISTRA and FRN.

Gibbons, M., Limoges, C., Nowotny, H., Schwarzman, S., Scott, P. and Trow, M. (1994) *The New Production of Knowledge: The Dynamics of Science and Research in Contemporary Societies*. Thousand Oaks, CA: Sage.

Gray, M. and Schubert, L. (2010) 'Turning base metal into gold: Transmuting art, practice, research and experience into knowledge', *British Journal of Social Work*, 40(7): 2308–25

Handmer, J. and Haynes, K. (eds) (2008) *Community Bushfire Safety.* Melbourne: CSIRO Publishing.

Harrisson, F. (2012) 'Digging the dirt: Locating design learning between people, politics and place', in M. Dodd, F. Harrisson and E. Charlesworth (eds), *Live Projects: Designing with People.* Melbourne: RMIT Press, pp. 153–66.

Higher Education Academy (HEA) (2012) *Education for Sustainable Development.* Accessed November 28, 2012 at: www.heacademy.ac.uk/education-for-sustainable-development.

Horne, R. and Fudge, C. (2014) 'International perspectives: Low carbon urban Australia in a time of transition', in C. Miller and L. Orchard (eds), *Australian Public Policy: The Challenges in Turbulent Times.* Cambridge: Policy Press.

Institute of Strategy and Policy on Natural Resources and Environment (IPONRE) (2009) *Viet Nam Assessment Report on Climate Change.* Accessed May 18, 2011 at: www.unep.org/pdf/dtie/VTN_ASS_REP_CC.pdf.

Kellogg Commission on the Future of State and Land-Grant Universities (1999) *Returning to our Roots.* Washington, DC: National Association of State Universities and Land-Grant Colleges,

Lave, J. and Wenger, E. (1991) *Situated Learning: Legitimate Peripheral Participation.* Cambridge: University of Cambridge Press.

Moback, U., Morrison, G., Knutson, P., Roth, S., Rhorn, P., Buhr, K. and Areslatt, H. (2012) *MISTRA Urban Futures: Reflections on the Transdisciplinary Process,* MISTRA Urban Futures Working Paper 2012: 2. Göteborg: Mistra Urban Futures.

Mulligan, M. and Nadarajah, Y. (2008) 'Working on the sustainability of local communities with a "community-engaged" research methodology', *Local Environment,* 13(2): 81–94.

——(2012) *Rebuilding Communities after Disasters: Lessons from the Tsunami Disaster in Sri Lanka and India.* New Delhi: Routledge.

Nolmark, N. (ed.) (2009) *Urban Knowledge Arenas—Re-thinking Urban Knowledge and Innovation.* Gothenburg: Final Report of COST Action C20, Urban Laboratory.

Perry, B. and May, T. (2010) 'Urban knowledge exchange: devilish dichotomies and active intermediation', *International Journal of Knowledge-Based Development,* 1(1/2): 6–24.

Polk, M., Kain, J.-H. and Holmberg, J. (2012) *MISTRA Urban Futures: A Living Laboratory for Urban Transformations.* MISTRA Urban Futures Working Paper 2012: 1. Göteborg: Mistra Urban Futures.

RMIT Architecture (2012) *Building the Community Projects, Vietnam, 2007–2012.* Accessed December 2, 2012 at: http://architecture.rmit.edu.au/Projects/building_the_community_vietnam.php.

Ryle, G. (1949) *The Concept of Mind.* London: Hutchinson.

Sara, R. (2006) *Live Project Good Practice: A Guide for the Implementation of Live Projects.* CEBE Briefing Guide Series, No. 8. Plymouth: University of Plymouth.

Sinatra, J. and Murphy, P. (1997) *Landscape for Health: Settlement Planning and Development for Better Health in Rural and Remote Indigenous Australia.* Melbourne: RMIT Press.

Spanier, G. B. (1997) 'Enhancing the capacity for outreach', *Journal of Public Service and Outreach,* 2(3): 7–11.

UNISDR (2011) *At the Crossroads: Climate Change Adaptation and Disaster Risk Reduction in Asia and the Pacific.* Accessed November 21, 2012 at: www.unisdr.org/we/inform/publications/21414.

van Schaik, L. (2005) *Mastering Architecture: Becoming a Creative Innovator in Practice.* Chichester: Wiley.

Whittaker, J. and Handmer, J. (2010) 'Community bushfire safety: A review of post-Black Saturday research', *Australian Journal of Emergency Management*, 25(4): 7–13.

7 Formative evaluation at a transdisciplinary research center

Tomas Hellström

Recent years have seen a surge in the establishment of Centers of Excellence as instruments for bringing together stakeholders around pertinent social topics, such as "Grand Challenge" type issues including sustainability, health, risk, and security, as well as similar cross-sector and strategic matters (Hellström 2013). These centers are usually interdisciplinary and deal with complex social and scientific challenges. They also run over a longer period of time than the typical research project (from 5–15 years), and they often involve a considerably broader spectrum of stakeholders than the usual research project. In other words, they typically are "radically cross-sector" (Orr *et al.* 2011; Aksnes *et al.* 2012).

In many ways, the Mistra Urban Futures center exemplifies this type of Center of Excellence. It was founded to deal with a specific issue—the challenge of urban sustainability—that can be described as a "wicked problem" in the sense of being so complex that it defies traditional disciplinary and policy boundaries (Rittel and Webber 1973). As a research topic, urban sustainability spans the social, engineering, and planning sciences, as well as drawing upon insights from other areas of research (urban sociology, political science, urban planning and design, ecology, and natural resource management). It also defies the social and political boundaries usually recognized in public policy, such as transport policy, waste management, housing policy, energy management, and so on. Developing knowledge about the conditions for urban sustainability involves problems in terms of the social coordination and integration of agendas and extant practitioners' knowledge, as well as in terms of bringing together the many modes of knowledge creation present in relevant academic fields. As noted in Chapter 1, this type of research has been referred to as "transdisciplinary" (cf. Gibbons *et al.* 1994), meaning that it spans several sectors and relates to practical rather than academic problems. It is a form of "co-production" of knowledge (Robinson and Tansey 2006) because it involves both academics and practitioners in the research process.

As has been illustrated for specific cities in Chapters 2–6, a center charged with this type of task faces a number of concrete challenges in attempting to secure socially and academically robust knowledge:

- Its participants must be able to establish coherent and systematic processes for involving stakeholders in knowledge creation. Furthermore, the knowledge so created must be relevant to center goals, amenable to being governed through center processes, and able to actually serve as a basis for sustainable change.
- Such processes must facilitate learning about transdisciplinary knowledge-creation processes, ultimately enabling them to be institutionalized without creating rigid structures and epistemic regimes. This means understanding the specific circumstances that condition transdisciplinary research and co-production of knowledge and adapting the organization accordingly.
- The complexity of the processes involved here must be balanced against the need for public accountability in the sense that even a center such as Mistra Urban Futures that is tasked with studying complex societal problems must be able to represent its inner and outer processes with a measure of simplicity. This means that the relation between internal center processes and those external processes taking place mainly in society (or "in the context of application," to use a common phrase) must be amenable to observation; i.e. they must be legible.
- Finally, it is now widely recognized that due to the long-term commitment of funders and other stakeholders, the lack of "off-the-shelf" or ready-made solutions, and the complexity of knowledge exchange and application, it is very difficult to evaluate such centers in terms of concrete impacts. Their outcomes are diffuse and uncertain, and they must be evaluated in terms of their processes and activities instead (Hellström 2011). Formative or process evaluations are appropriate in cases where the focus is on what stakeholders actually do, and what they themselves perceive as leading to positive change.

This chapter outlines some of the challenges involved in managing and evaluating quality at a transdisciplinary research center such as Mistra Urban Futures, and sets out to delineate a number of principles for handling the challenges that emerged during the course of initiating and running the Mistra Urban Futures center. The chapter begins by identifying some relevant previous research on quality and evaluation in inter- and transdisciplinary research as well as previous work on the quality of research centers and Centers of Excellence. It then provides an account of lessons learned from Mistra Urban Futures, as well as presenting and illustrating the application of a formative evaluation framework developed in the course of establishing this center. The chapter ends with a concrete suggestion for how to continue to develop an evaluation framework for complex and diffuse research based on eliciting actors' own understandings of how their actions transform into valued effects.

Evaluating transdisciplinary research centers

The evaluation of transdisciplinary or co-produced research poses a great many challenges to traditional evaluation. First, there is the question of how quality should be assessed in this new type of hybrid science (Hemlin and Rasmussen 2006). There are several traditional quality markers in science that researchers regularly apply in evaluating hypotheses, experimental set-ups, and lines of argument as part of their everyday practice—for example when a colleague argues the significance of this or that result to the development of theory. Researchers in philosophy and methodology of science have proposed that dimensions of value such as simplicity, scope, precision, and fruitfulness are appropriate for this kind of internal scientific evaluation (e.g. Kuhn 1979). These quality markers have to do with evaluating theory. However more important to our purposes are perhaps those aspects of scientific conduct that promote such theories' "social and ecological validity". This involves generalizability or "transferability" to new problem contexts, credibility of the experts involved vis-à-vis stakeholders, and a broadened conception of what is "evidence" and legitimate channels for research communication given unclear or moving problem boundaries.

As research proposals and results are increasingly being judged from the point of view of their social and economic benefits challenges such as these typically come to the fore. This involves evaluating arguments of "social relevance" advanced by center stakeholders, with evaluators sometimes attempting to project social impacts such as technical and social innovations, or industrial process improvements. Key issues here include: How should trade-offs between scientific and social values be handled? How do we decide on policy and criteria in this regard? And how do we operationalize an evaluation scheme that captures precisely those values we want to emphasize? In such cases it is not enough to develop "checklists" and tick-off boxes; rather one has to look at the entire system of knowledge governance developed by a center, and try to understand how these processes and activities can promote knowledge and action in the area of concern. This is equally a form of organizational analysis as it is an assessment of the quality of knowledge creation. We will take a closer look at this challenge by reviewing a few statements on how inter- and transdisciplinary research ought to be evaluated.

A common definition of *inter*disciplinarity is that it is a form of scientific inquiry that integrates two or more disciplines in order to solve a scientific problem. Unlike *multi*disciplinarity, where disciplines can work side by side and independently on a common topic, interdisciplinarity implies that the disciplines adapt to their new partners, and in so doing form a new line of inquiry. *Trans*disciplinarity, on the other hand, combines this mode with the further complication of also integrating practitioners into the process of inquiry, and applying the research effort to an area of practical concern (see Hessels and van Lente 2008 for an overview of this discussion). The assumption here is that some social problems can only be addressed by transgressing disciplinary and sector boundaries. These include issues of sustainability or risk—topics that cannot be delineated or addressed by the scientific community alone. The more radical vision, that of transdisciplinarity, identifies this new approach to research as "Mode 2", as opposed to traditional,

Table 7.1 Mode 1 and Mode 2 research (adapted from Thorpe and Beasley 2004)

Dimension	*Mode 1*	*Mode 2*
Knowledge context	Academic community	Application context
Nature of knowledge	Disciplinary	Transdisciplinary
Producers	Homogeneous	Heterogeneous
Quality control	Peer review	Social, economic, reflexive
Creativity	Individual phenomenon	Group phenomenon

discipline-oriented basic science, referred to as "Mode 1", (Gibbons *et al.* 1994). Table 7.1 summarizes the distinction between traditional modes of academic research (Mode 1), and transdisciplinarity (Mode 2), in terms of their context of knowledge production, the nature of this knowledge, the composition of the community of knowledge producers, the type of control exercised over processes, quality control, and how creativity is perceived. It is clear from this comparison that the nature of knowledge production in Mode 2 is much more complex, socially inclusive, open-ended, and hard to manage than that of traditional academic research in Mode 1.

Inter- and transdisciplinary research exist precisely because they are needed to solve complex problems. In the first case, the frontier of advanced strategic science is interdisciplinary more often than not, and social problems, such as Grand Challenges, typically require the transdisciplinary approach of Mode 2 research. The research commissioning system, however, is usually ill equipped to assess the value of research of this kind, as its most prominent methods—peer review and metrics—seem to be thoroughly discipline-oriented or rooted in Mode 1 (Bergmann *et al.* 2005; OECD 2009). Developing methods for assessing the quality of this type of science is not only a central concern of traditional funding bodies and of sector research funding actors; it should also be a concern for center managers and stakeholders, who otherwise may find themselves short-changed by unrealistic, traditional evaluation models. However even though transdisciplinary, co-produced research often addresses local concerns, evaluation standards must still be rigorous and comparable across funders and projects, and must be anchored in a solid understanding of what quality is, and what it can be in such cases. One of the most prominent methods for assessing inter- and transdisciplinary research is the use of mixed evaluator groups—for interdisciplinarity, members should cut across disciplines, and for transdisciplinarity, they should also cut across sectors (aka "user involvement" or "extended peer review," see, for example Funtowicz and Ravetz 1993). Interim and process evaluations put greater emphasis on how projects and centers go about integrating stakeholders in formulating and executing inquiries, while post-hoc, summative evaluations may look more at social impacts (e.g. among stakeholders) and impacts across the disciplines involved in the project (Walter *et al.* 2007). Evaluation questions that one would typically find in these cases, but which would be absent from traditional discipline-based evaluation, concern issues such as (Pohl 2011):

- How diverse are the participating disciplines, methods, theories, and social actors involved? Does the combination map onto a real problem in a novel and relevant way?
- What is the approach to stakeholder integration used in the project?
- How are self-reflection and adaptation to the unfolding problem context maintained in the project?
- How likely is the project to make a substantial contribution to social problem solving?
- Can one assess the participants' collaborative skills? Is management prepared to facilitate collaboration?

Huutoniemi (2010) notes that the most common approach in evaluating interdisciplinary research is to apply the criteria used for disciplinary fields, in many cases those criteria that exist in the fields involved in the interdisciplinary effort, e.g. parsimony, simplicity, scope, etc. She argues for using other values instead, such as integration and synergy, which emphasize how disciplines and social interests are integrated into a new knowledge product. In this case, quality is about how deep and relevant this integration is, from the point of view of the problems that the project sets out to address. Criteria such as cooperation and cross-fertilization between fields are the most important, and often the focus is on process criteria such as communication, learning, and integrative activities in the project or center. One problem with this approach is the risk of singling out inter- and transdisciplinarity as a "special case" next to traditional research, and inadvertently marginalizing it. Another issue is how to identify truly interdisciplinary projects before they have started, given that many of the criteria have to do with processes applied in running them.

A more radical approach would be to jettison traditional epistemic values altogether and replace them with social (e.g. democratic) values that relate to political and social needs, and for research to be evaluated using such values, for example with the help of lay expertise (evaluators from outside the academic community). The problem with this approach is that it does not show how, if such criteria were applied, science could avoid transforming into something else, for example social services provision. It leaves open a crucial question which needs to be resolved in praxis: When should scientific values guide evaluation and when should social values do so? (Huutoniemi, 2010).

Examining the conditions for this type of evaluation, Klein (2008) sees the ability to manage different types of experts (or expertise) as an important factor for success. This involves calibrating individual and disciplinary standards (as well as standards that cut across sectors in society, in the case of transdisciplinary research), and constructing new "cultures of evidence" that can handle the complexity and tradeoffs that exist in transdisciplinary projects. Seven principles, or areas, that need "management" are singled out as especially important for accomplishing this:

1 *The variability of goals*: This means that the goals of interdisciplinary research are usually more flexible than those of traditional research, and that the creation of artifacts and other social, pragmatic goals usually requires more inter- or transdisciplinarity.
2 *The variability of criteria and indicators*: Inter- or transdisciplinarity usually requires less use of proxy indicators (such as citations as proxies for scientific quality) and more emphasis on "intrinsic indicators," such as what counts as "good work" given the problems at hand. This can also include aesthetic qualities and problem-solving capacity.
3 *Leveraging integration*: This has to do with how well the constituent perspectives/fields are integrated in the project. For example, have the right partners been selected, are structures flexible enough to further integration, and is there a common vision or framework to keep actors together?
4 *Interaction of social and cognitive factors in collaboration*: This category resembles the previous one, but stresses how well the social practices in the group or project manage to promote intellectual integration and joint problem solving. Are there, for instance, joint activities that help participants across the organization negotiate their perspectives?
5 *Management, leadership, and coaching*: Integration should be carefully worked into an organization's management structure. Both the research and the evaluation process must be coached to encourage repeated discussions among researchers and between evaluators and project applicants as a way to promote mutual learning and understanding of each other's perspectives.
6 *Iteration in a comprehensive and transparent system*: This has to do with constantly testing and communicating results and ideas across the project's stakeholder groups, inside and outside of the center, and feeding the results of this interactive process back into the project's goals.
7 *Effectiveness and impact*: This is the recognition that inter- and transdisciplinary projects have impacts across fields of praxis and academic fields, and must be assessed accordingly, e.g. in terms of how external target groups make use of the knowledge, and how particular results or methods can have various impacts depending on how they diffuse into sectors and disciplines.

The previous accounts make it clear that evaluation and quality processes for transdisciplinary research must take account of a mix of epistemic and social criteria for quality in such a way that: (1) There is a sensitivity to how quality is assessed in terms of the uniqueness and variability of goals among several stakeholders; (2) Social mechanisms and structures such as adequate leadership are in place to secure the robustness of processes and the integration of multiple stakeholders and perspectives; and (3) The impacts of research are captured adequately. This mix of dimensions must be represented in a viable organizational form that can be evaluated in lieu of concrete social impacts, which themselves may not materialize for some time.

In recent years multi-, inter-, and transdisciplinary research centers, sometimes referred to as Centers of Excellence, have become a significant means of channeling funding for research in many countries. This approach to funding basically requires that a center be formed and a sufficient number of researchers be brought together around common topic. Often this topic is of strategic concern, and involves fundamental research challenges as well as substantial promise in terms of technological and/or socio-economic returns (Aksnes *et al.* 2012). An equally central characteristic of the Center of Excellence Concept is that high international standards, or "scientific excellence," are applied in evaluating proposals for such centers and their operation. It is precisely the mix of aims, such as scientific excellence, in terms of international recognition in the field; technological progress, in terms of transferring useful knowledge to the business community; and other socio-economic outcomes, that typifies these collaborative research environments (Hellström 2013). As we have seen above this hybridity is also a source of challenges for evaluation.

One aspect of investing in centers rather than in projects is that the former are more long term, and usually considerably better funded than a project. This enables centers to employ many more researchers than single projects, which often may be necessary to achieve the spread among stakeholders needed to tackle complex problems (Orr 2011). From these premises a number of evaluation challenges follow. For one, it is important for funders, given the amount of money and the time involved, to develop and apply interim evaluation instruments so that they can track a center's progress over time. This can be measured in terms of academic production, for example in the form of journal articles, but also in relation to the organizational viability of the center. Unlike a small project, the success of an organization of say 30 people is critically dependent on having viable structures and processes (Rip 2003). Interim organizational (or process) evaluations have increased in importance as a research evaluation methodology mainly due to the emergence of Centers of Excellence.

These centers are also often expected to generate impacts in the broader community. This has led to other types of evaluation concerns, for instance attempts to measure the capacities in the science and policy system that research centers are expected to generate. These include capacities pertaining to research leadership and organization, priority setting, attraction of human capital, industrial and policy strength in certain fields, and socio-economic impacts such as social problem solving related to national challenges (e.g. disease mitigation, agricultural improvement, and others). Herein lies a challenge of assessing impacts, but also of disentangling viable from spurious arguments about such impacts as they appear in application texts and center reports. Specific issues include how to construct indicators for organizational viability, and in what dimensions impacts can be expected to materialize from such centers (Hellström 2011).

Drawing on case studies from the natural sciences Heinze *et al.* (2009) identify a number of drivers for creativity in these types of organizations and outline them in terms of process dimensions of quality. The study finds that research quality is associated with small group size, organizational access to a complementary variety of technical skills, stable sponsorship, access to extramural skills and resources, and facilitating leadership, and that all of these factors are dependent on some contextual aspects connected to the organizational and institutional environment. A number of specific points are brought to the fore, e.g.:

- Freedom to pursue problems within a larger set of constraints (what can be referred to as "carefully constrained autonomy");
- Group size that fosters close student–mentor relationships;
- Mechanisms that facilitate interdisciplinary contacts;
- Collaborative networks that facilitate access to complementary skills and knowledge;
- Active leaders who have a vision and are themselves practicing researchers;
- Flexible and stable research funding arrangements;
- Job mobility, hosting of visitors, and visiting other sites of research;
- Friendly competition among participants.

To a large extent these dimensions refer to "internal" aspects of center operations. While these are no doubt important, transdisciplinary research also depends on the ability to integrate stakeholder knowledge and practical concerns into the research process. This may require a more multidimensional and inclusive idea of quality. In a recent article, Hellström (2011) analyzes a number of Swedish Centers' of Excellence mid-term evaluations in order to identify common themes and priorities applied by the evaluators of these centers. Here excellence is seen as a multidimensional construct. It can include the quality and quantity of the scientific output of the center in terms of research results, as well as the quality of center processes such as strategies for recruitment, organization, and integration with international scholarship and local university departments, as well as with society at large. In addition, progressive leadership is a key dimension for such centers (see Hollingsworth and Hollingsworth 2000; Pfirman and Martin 2010), as is having center activities with the potential to impact on society at large, for example through innovation and social improvement.

Hellström (2011) identifies seven evaluative dimensions employed by evaluators to assess center processes:

- *Organizational capability* has to do with the center's ability to enable its members using organizational solutions, and to respond rapidly and effect cooperation.

- *Strategic capacity* refers to the center having accurate situational awareness and the ability to employ leadership to make the right moves in a changing world.
- *Resource strategy* concerns how the center acquires external resources and allocates these internally to achieve certain ends.
- *External cooperation and networking* has to do with creating high quality partnerships across sectors and sustaining these over time.
- *Human capital base* relates to the center's recruitment policy, access to crucial talent, and quality of leadership.
- *Epistemic capacity* concerns the center's intellectual coherence, access to new knowledge, learning mechanisms, and integration of disciplines.
- Finally, the *research-teaching nexus* refers to how successfully the center shares its learning with students and cooperates with education.

When it comes to evaluation, one size does not fit all. The relative emphasis on the factors discussed above needs to be tailored to the aims and structure of the center in question. It is important to conceive a center's internal and external processes within the same framework and to try to harmonize formal center processes with those informal and messy external processes that constitute the "context of application" or the actual arena of knowledge co-production. In what follows we will look more closely at a framework for quality and evaluation management developed at Mistra Urban Futures. We will introduce this account with a discussion of why formative process evaluations are important for such a center.

Lessons learned from Mistra Urban Futures

There are several reasons why Mistra Urban Futures engages in evaluation of center activities and processes. First, such evaluation collects information about activities and highlights the actions and effects of the center more clearly for its participants and stakeholders. It serves a communicative function. Second, evaluation makes it possible to assess the strengths and weaknesses of the activities and structures of the center in relation to center goals, and informs discussion on how they can be achieved. In this sense, it fills a performance monitoring function that also provides opportunities to assess performance in relation to other similar centers, i.e. to benchmark. Third, evaluation creates an opportunity to track the effects of center activities, including tangible and intangible outcomes, allowing leadership the opportunity to intervene where necessary and on the basis of what the center is accountable for to its stakeholders. This may be referred to as a governance function. An important fourth function is the process by which quality management and evaluation enable the center to develop and improve by making members aware of the significance of planned and unplanned center activities and their effects. This is the learning function of the formative

Table 7.2 Formative evaluation functions

Formative evaluation functions	*Outcomes*
Communication, internal and external	Situational awareness of center participants and stakeholders regarding progress and results.
Performance monitoring and benchmarking	Identifying and adhering to relevant and adequate measures of excellence in the field.
Leadership and governance	Enabling center principals (board, funders, etc.), management, and participants to direct and influence activities appropriately.
Continuous improvement and learning	Acting as a knowledge hub on center structure and best practices thereby ensuring that past successes and mistakes are taken into account in developing the center.
Legitimacy	Communicating how and to what extent center goals are achieved, so as to enable internal and external assessment of the level of quality vis-à-vis expectations.

evaluation activity. Finally, there is the question of legitimacy. Here, the task of the evaluation activity is to develop center processes in the direction of legitimate stakeholder expectations. These functions are summarized in Table 7.2.

The complexity of the functions and goals outlined above suggests that formative evaluation should evolve continuously with the development of the center. That is, it should be sensitive to emerging conditions and the center's aims. It is especially important that it is not an isolated function, but is integrated into center life in general, and viewed as an integral and supporting part of the research and communication in the center sub-projects and its networks. This means that formative evaluation should be useful to participants' changing agendas, and not devolve into the kind of rigid framework observed in some R&D organizations where accounting and monitoring systems lead to "information or data collection paralysis," goal displacement, and misuse (cf. Feller 2002).

This is best avoided through a framework that balances several types of center dimensions. Some of these will certainly be of an operational character, documenting and feeding back information for center management and evaluation purposes, while a second set of dimensions, closely associated with the first, is more about critically assessing the value and ideas underlying activities and aims. This is about facilitating learning, so that goals and approaches can be revised accordingly.

In this way two key dimensions of center management emerge: one that is operational or administrative, and deals with the everyday handling of the center; and one that is learning-oriented and self-reflexive and deals with more fundamental conditions for learning and co-production at the center.

Table 7.3 Operational and learning dimensions of center management

	Operational	*Learning*
Internal	(a) Local output/outcome indicators and communication	(c) Improving center theory by mapping values and participant accounts
External	(b) Network monitoring and quality communication	(d) Knowledge of value creation for transdisciplinary research

Both these dimensions are needed to fulfill the demands on evaluation elaborated above by Klein (2008), Huutoniemi (2010), and Hellström (2011).

However, in addition to these two dimensions—operational and learning—the center also exists in relation to a world outside, namely its stakeholders, networks, and cooperative partners. A center that aspires to have an impact outside of its own organization must work on these operational and learning goals internally as well as externally, which adds two more dimensions to the mix: an internal and an external. In this way we end up with four dimensions that are needed for management and evaluation of a transdisciplinary center (see Table 7.3).

In what follows, these dimensions will be explained and further developed in relation to some of the five formative evaluation functions presented in Table 7.2.

Operational internal–local center governance

This includes identifying/monitoring effects of the center in academic and practical/societal terms. An essential goal of academic quality work in this case is the identification of concrete outputs/outcomes of the center, and the use of this information by participants for governance purposes. Some of these outputs/outcomes are self-evident; in the academic sphere, for example, they include scholarly outputs such as publications and citations, while in the context of application they may include new products/services, decisions, policies, and improvements experienced by stakeholders (Walter *et al.* 2007). However in a transdisciplinary setting it is important to identify new and non-obvious consequences as well as intermediary effects of the center activities that have the potential to generate impacts in the long run (Balderston 1995). One of the functions of internal center governance is to develop criteria and indicator sets for such impacts and apply them in the center's operational processes. This involves updating and operationalizing traditional quality criteria employed for academic as well as policy work (Boix Mansilla 2006).

Operational external–network governance

The governance of a center usually involves a relationship between on the one hand some kind of governing board and a management team, and on the other hand the management activities connected to the consortium partners, center funders, regional, academic and business actors, and other stakeholders. While the former is an example of a traditional governance relationship related mainly to operational aspects of a center, the latter represents network governance (Newman 2004). Network activities of this type can be subject to evaluation and quality monitoring, including the provision of a strategic framework and an operational indicator set for network management and governance. Here the challenge is to identify appropriate principles for these interactions as well as to develop and apply operational indicators to collect and feed back information on these relationships and their relevant effects.

As suggested above an important second area for evaluation is to help identify and make visible the values guiding a center, such as its overarching goals, in order to facilitate learning, as well as to contribute an evidence-based platform for continuously moving closer to a center's value ambitions (for example sustainability and transdisciplinary knowledge creation). This involves two main activities:

Learning internal–explicating how the creation of social and academic value is perceived

This includes mapping participants' conceptions of how value is created at the center and on the basis of this formulating a normative framework for how activities, outputs, and outcomes are to be valued. On the organizational/process level this means explicating the values and associated processes for facilitating transdisciplinary and cross-sector, application-driven knowledge creation (co-production) and identifying the relevant outputs (what concretely is valued in these processes and how). This also involves translating the values underpinning the center (including notions such as Sustainable Development in general and Sustainable Urban Development in particular) into terms that make it possible to identify valued impacts and activities. The value framework is an essential governance mechanism and serves to identify and promote as relevant certain activities rather than others, as well as to formulate novel and relevant evaluation indicators vis-à-vis external stakeholders (Hellström 2011).

Learning external–meta-function: understanding transdisciplinary value creation

Evaluation is one of the main ways for a center to collect insights about internal and external knowledge-generating processes. One way of doing this is by mapping, documenting (including critical assessment via the academic literature), and providing feedback on transdisciplinary work processes from an evaluation perspective—that is, to use evaluation to create knowledge about value creation processes inside and outside the center. The knowledge generated in this way will contribute to center goals by representing the center as a unique case from which best practice can be transferred to other settings;

the audience thus consists not only of local center stakeholders, but of the transdisciplinary and sustainable urban development research community at large.

In what follows we will look more closely at a framework for implementing these evaluation functions, one that captures the center participants' own appreciation of how actions and activities are connected to internal and external value creation. We refer to this as an "action–value attribution framework." It represents a "theory-in-use" or a cognitive map of the center's value-creation processes for purposes of evaluation (see Hodgkinson *et al.* 2004; Weiss 1997). These notions will be further explained below.

Understanding the center theory in-use: the "action–value attribution framework"

The above framework for formative center evaluation emphasizes two dimensions of evaluations, namely internal–external and operational–learning aspects. Value is usually created when these dimensions interact and evolve together over time, for example when the boundaries between internal and external activities change, and new learning from external partners leads to changes in the center's internal management.

The problem here is that how a given center generates these values is highly dependent on context and on factors such as:

- the center's aims;
- the actors involved;
- their composition;
- the policy context; and
- the terms of reference.

All these and probably many more aspects affect what actions are appropriate for furthering center processes and goals. This suggests that pre-given standard frameworks will probably not apply very well (at least not during formative evaluations, where the precursors to success must be identified, evaluated, and promoted). Rather, it is imperative to derive a locally valid evaluation framework from the involved actors' own appreciation of their aims, and their conditions for realizing those aims. A formalization of these aims and conditions represents the "theory-in-use" of a given center (Argyris and Schön 1996). A theory-in-use is basically an actor's tacit framework for structuring and solving problems and understanding situations, and can be extracted from an actor using interviews and systematic analysis. Such analysis can be used to create a representation or model of what activities and conditions are perceived as appropriate and of what outcomes are sought from such activities. This is what we refer to as an "action–value attribution framework," namely a framework showing how a center participant attributes value to activities and their effects.

For a formative evaluation to be successful it should be constructed on top of such a framework; that is, it should use the center participants' theories-in-use and a derived map of their value attributions to construct indicators for success. The result is an "evaluation theory" for the center based on an understanding of how actors perceive value creation—how they attribute value to actions in terms of outcomes or intermediary outputs. Mapping these understandings represents an important form of learning for the center, in addition to being useful for evaluating outcomes and effects of the center and assessing the viability of its operational processes. It involves short- and long-term mapping of the transdisciplinary processes and effects for the purpose of learning *and* evaluation.

In the following we will look at one methodology for this process, and suggest how it can be used to realize the ambitions of formative center evaluation set out here. As described above this activity involves capturing the center participants' values, in terms of how key contributions, activities, outputs, and expected as well as observed outcomes are assessed and judged as contributing to the center goals. The aim of this exercise is to develop evaluation theory by mapping such values and accounts, thereby not only providing a basis for critical and constructive reflection on the subject matter and on the center itself, but also (and importantly) to identify non-obvious evaluative dimensions of the center. In order to do this several interviews were conducted with the project leaders in charge of five projects during the center's start-up period. These were one-year pilot projects undertaken to establish a base-line methodology or approach to transdisciplinary co-production, as well as to try out a number of interesting thematic leads for the center's Gothenburg operations. As mentioned in Chapter 2, the interviews were conducted with the project leaders (one academic and one practitioner) in the initiation, starting, mid-term and end phases of their respective one-year projects. The final interview aimed at eliciting responses about conditions, aims, and perceived success factors of the projects, as well as of the center as a whole (cf. Gugiu and Rodriguez-Campos 2007), and included the following questions:

- What activities are considered central to achieving your objectives?
- What are the direct and indirect results of these?
- What conditions in society do you attempt to create with your sub-projects?
- Give examples of effects that your project has the best possibilities to bring about?
- What conditions and what actors affect your project the most, directly and indirectly?

The interviews were recorded and transcribed. The transcripts were analyzed and statements were identified where interviewees made direct reference to problems and opportunities for their projects and their stakeholders, activities they conducted or perceived to address these issues, and other valued outcomes of their research. This analysis followed the "general inductive approach for analysis of qualitative evaluation data" developed by Thomas (2006). The results of this analysis, in a summary form, were a number of action–effect relationships (shown below). These capture what project leaders expressed in the interviews in terms of what activities or conditions could lead to certain valued effects. This is what we refer to as an action–value attribution framework. Three main areas of such attributions were identified: *professional cooperation*, *cognitive integration*, and *resource provision*. In addition, under each of these areas a number of sub-themes were observed, referring to activities or states perceived to generate certain valued outcomes. These are summarized below (Figure 7.1). The boxes on the left refer to types of activities and conditions that are seen as having positive effects, the arrows indicate attributed influence, and the boxes to the right summarize the outputs and effects that these activities are perceived as leading to. The formative evaluation takes the right-hand outcomes as more or less desired (depending on what are perceived as aims, what is lacking, etc.) and evaluates the conduct of left-side activities accordingly.

These results suggest that when conducting formative process evaluations of a transdisciplinary center, like Mistra Urban Futures, the assessment should focus on:

- Issues concerning cooperation across professional and sector divides.
- How well networks are made to integrate/overlap through, for example, meeting routines.
- How well various competences are combined through task integration and division of labor.
- Activities aiming to create joint framings of the goals and methods applied.
- Resource provision, which includes allocating adequate time and relevant personnel.

If such conditions are fulfilled and the activities are performed, certain valued outcomes are perceived to come about, such as project effectiveness, attraction or resources, relevance of results, learning, and new projects being relevant to stakeholders. This framework is not, of course, a complete methodology for evaluation, but represents a tentative model for how to focus process evaluation in order to develop the center. It is tentative in the sense that the very same context dependence that makes this framework more valid than a standardized one, also threatens to make it obsolete unless it is regularly updated.

The next step in this process is to elaborate a methodology, including indicators or at least questions, to conduct process evaluations of the center. It is also important to develop routines for updating the framework and communicating the insights from the formative evaluation to stakeholders.

Activities		**Valued effects**
Professional Cooperation		
New cooperative constellations • Public sector–business cooperation • Practitioner–academic cooperation	⇨	• New projects and project applications • Cross-sector networks and social capital
Network integration • Overlapping social and professional networks and working history • Cooperation and meetings between projects	⇨	• Project effectiveness • Coordination vis-a-vis ongoing political dialogues/activities • Resource provision
Cognitive integration		
New knowledge combinations • Competence-based division of labor between practitioners and academics • New combinations of competences and disciplines • Thematically overlapping tasks between projects and other professional activities	⇨	• Initiation of new projects • Time to realize the project • Varied and relevant results
Common points of departure • Joint project initiation • Joint discussion of methods • Sharing knowledge of relevant stakeholders and their needs • Coordination of research process with political, administrative, budgetary decisions	⇨	• Effective project build-up • Project learning • Adapted academic output • Responsive stakeholders • Resource attraction
Resource provision		
Time availability • Time availability for empirical research and data analysis and discussion • Time to discuss across sectoral/professional divides	⇨	• Robust and relevant results • Adaptation and extension of results to application context • Possibility to address a larger number of questions. • Grasp of the "full picture" → citizen perspective

Figure 7.1 An action–value attribution framework for Mistra Urban Futures

A way forward

The evaluation model developed here can be made more comprehensive and also be made to involve stakeholders in a more systematic fashion. Such an "extended" formative evaluation process could proceed in the following fashion:

1 In a first phase, a value-mapping activity can be designed that uses facilitative interviews, praxis reflection, and focus groups to elicit values guiding the center's participants with regard to the subject matter.
2 Alongside (1), or perhaps as part of it, a similar exercise focuses on activity mapping, with the aim to identify central value-producing activities, important interfaces with other center participants, and the content and assumed logic driving valued processes. Here one needs to capture, in structural terms, where critical interactions and value-producing activities in and around the center take place. This amounts to an "impact sampling strategy" for the center. It is probably best captured through visual representations of the center's structure, its surrounding network, and its processes, including indications and descriptions of where the value-producing activities are located.
3 The values and activity mapping (including valued, expected, and actual impacts) is used to generate a "living" online evaluation paradigm, which is unique to the center and which can facilitate center interaction, serve as a basis for adjusting, concretizing, and revising mutual expectations, and contribute new knowledge on transdisciplinary management. This representation should be updated regularly as part of the ongoing collection of new data and participant feedback.

Conclusion

The aim of this chapter was to capture some of the challenges involved in managing quality—in terms of formative process evaluation—at a transdisciplinary research center. It set out to delineate a number of principles for evaluation and quality management as presented in the literature as well as several of the challenges that emerged during the course of initiating and running Mistra Urban Futures. An important insight from the overview of perspectives on transdisciplinarity and interdisciplinary centers was that evaluation needs to be tailored to the aims and structure of the center in question. One challenge is to create a productive connection between the routines and activities of a center, its processes and operations, and the ways the center enables learning and co-produces knowledge with its stakeholders in the context of application.

The introductory chapter argued the need for transgressing the linear model of knowledge production and policy intervention. This means to stop thinking of the relevance of knowledge in traditional terms such as "reliability," and instead ask for the "social validity" or "robustness" of knowledge. The present chapter sought to address this challenge by asking how we might go about validating knowledge and knowledge production as "good" in this more extended fashion. What we have found is that there are in fact ways in which an evaluation process can accomplish this, for example by:

- bridging the boundaries between inside and outside of a center;
- making sure that stakeholders are involved in mutual learning with academics;
- representing the complexity of urban problems (and of co-production) without losing the simplicity needed for accountability; and
- providing accounts of "impacts" even when the journey is clearly nowhere near finished.

Concepts such as "transdisciplinarity" and "co-production" are helpful here. However their greatest utility does not lie in their ability to provide distinct models for "how to do it," but rather to provide new aims for researchers and practitioners and new standards for interaction. These aims and standards form the basis for evaluating existing policy frameworks and suggesting new ones.

One of the most important consequences of these "refocusing devices" is the way in which they help us see that knowledge is carried by a variety of actors, and that expertise is not an academic monopoly, but rather distributed in networks and activated by encounters across sector boundaries. Another key insight, and a consequence of the previous one, is that excellence in this type of research cannot be defined according to a standard academic measurement stick, but that it depends on social relevance. Scientific excellence in a trans-scientific setting is achieved by means of being able to mirror, in the process of research and its organizational setting, the concerns and issues that motivate stakeholders to pursue similar inquiries, and to involve them in the process of creating new knowledge. An evaluation framework that satisfies these conditions must therefore be able to do several things at once:

1. It must be able to collect and process information of value not only from the organization itself but also from its extended and often messy context.
2. It must be able to capture and represent the value created by and for an extended and amorphous group of participants, whose ideals, beliefs and aims diverge.
3. It must do this in a way that is clear enough to make good on the need for accountability, that is to render this complexity in clear and understandable terms.

Some of the difficulties related to this can be addressed by applying an action–value attribution perspective as described above. Such a perspective connects activities and processes to center goals in a way that is endorsed and accepted by center participants; it enables learning in terms of the center's goals and conditions and their contingent dependence on context, and it accepts accountability without sacrificing relevance in the sense that what is "measured" is what stakeholders actually are doing, and themselves perceive as leading to positive change. Research and praxis regarding center evaluations is still a fledgling area, and approaches that utilize program theory thinking, as in the above case, are even more so. Hopefully the above discussion can make a contribution to this important area of research evaluation.

References

Aksnes, D., Benner, M., Brorstad Borlaug, S., Foss Hansen, H., Kallerud, E., Kristiansen, E., Langfeldt, L., Pelkonen, A. and Sivertsen, G. (2012) *Centers of Excellence in the Nordic Countries.* Oslo: NIFU.

Argyris, C. and Schön, D. (1996) *Organizational Learning II: Theory, Method and Practice.* Reading, MA: Addison Wesley.

Balderston, F.E. (1995) *Managing Today's University: Strategies for Viability, Change, and Excellence.* San Francisco, CA: Jossey-Bass.

Bergmann, M., Brohmann, B., Hoffman, E., Loibl, M.C., Rehaag, R., Schramm, E. and Voss, J.-P. (2005) *Quality Criteria of Transdisciplinary Research: A Guide to Formative Evaluation of Research Projects.* Frankfurt am Main: ISOE.

Boix Mansilla, V. (2006) 'Assessing expert interdisciplinary work at the frontier: An empirical exploration', *Research Evaluation*, 15: 17–29.

Feller, I. (2002) 'Performance management redux', *American Journal of Evaluation*, 23: 435–52.

Funtowicz, S.O. and Ravetz, J.R. (1993) 'Science for the Post-Normal Age', *Futures*, 25: 739–55.

Gibbons, M., Limoges, C., Nowotny, H., Schwartzman, S., Scott, P. and Trow, M. (1994) *The New Production of Knowledge: The Dynamics of Science and Research in Contemporary Societies.* London: Sage.

Gugiu, P.C. and Rodriguez-Campos, L. (2007) 'Semi-structured interview protocol for constructing logic models', *Evaluation and Program Planning*, 30: 339–50.

Heinze, T., Shapira, P., Rogers, J. D. and Senker, M. S. (2009) 'Organizational and institutional influences on creativity in scientific research', *Research Policy*, 38: 610–23.

Hellström, T. (2013) *Centers of Excellence as a Tool for Capacity Building.* Paris: OECD.

——(2011) 'Homing in on excellence: Dimensions of appraisal in Center of Excellence program evaluations', *Evaluation*, 19: 306–16.

Hemlin, S. and Barlebo Rasmussen, S. (2006) 'The shift in academic quality control', *Science, Technology and Human Values*, 31: 173–98.

Hessels, L. and van Lente, H. (2008) 'Re-thinking new knowledge production: A literature review and a research agenda', *Research Policy*, 37: 740–60.

Hodgkinson, G.P., Maule, J.A. and Bown, N.J. (2004) 'Causal cognitive mapping in the organizational strategy field: A comparison of alternative elicitation procedures', *Organizational Research Methods*, 7: 3–26.

Hollingsworth, R. and Hollingsworth, E. J. (2000) 'Major discoveries in biomedical research organizations: perspectives on interdisciplinarities, nurturing leadership, and integrated structures and cultures', in P. Weingart and N. Stehr (eds), *Practicing Interdisciplinarity.* Toronto: University of Toronto Press: 215–44.

Huutoniemi, K. (2010) 'Evaluating interdisciplinary research', in R. Frodeman, J. Thompson Klein and C. Mitcham (eds), *Oxford handbook of Interdisciplinarity*. Oxford: Oxford University Press: 309–20.

Klein, J.T. (2008) 'Evaluation of interdisciplinary and transdisciplinary research: A literature review', *American Journal of Preventive Medicine*, 35(2S): 116–23.

Kuhn, T. (1979) *The Essential Tension*. Chicago, IL: University of Chicago Press.

Newman, J. (2004) 'Constructing accountability: Network governance and managerial agency', *Public Policy and Administration*, 19: 17–33.

OECD (2009) *Enhancing Public Research Performance Through Evaluation, Impact Assessment and Priority Setting*. Paris: OECD.

Orr, D., Jeager, M. and Wespel, J. (2011) *New Forms for Public Research: A Concept Paper on Research Excellence Initiatives*. Paris: OECD.

Pfirman, S. and Martin, P.J.S. (2010) 'Facilitating interdisciplinary scholars', in R. Frodeman, J. Thompson, Klein and C. Mitcham (eds), *Oxford Handbook of Interdisciplinarity*. Oxford: Oxford University Press: 386–403.

Pohl, C. (2011) *Questions to Evaluate Inter- and Transdisciplinary Research Proposals*. Working paper, Berne: td-net for Transdisciplinary Research.

Rip, A. (2003) 'Societal challenges for R&D evaluation', in P. Shapira and S. Kuhlmann (eds), *Learning from Science and Technology Policy Evaluation*. Cheltenham: Edward Elgar Publishing, pp. 32–53.

Rittel, H.W.J. and Webber, M.M. (1973) 'Dilemmas in a general theory of planning', *Policy Issues*, 4: 155–69.

Robinson, J. and Tansey, J. (2006) 'Co-production, emergent properties and strong interactive social research: the Georgia Basin Futures Project', *Science and Public Policy*, 33: 151–60.

Thomas, D.R. (2006) 'A general inductive approach for analyzing qualitative evaluation data', *American Journal of Evaluation*, 27: 237–46.

Thorpe, R. and Beasley, T. (2004) 'The characteristics of performance management research: Implications and challenges', *International Journal of Productivity and Performance Management*, 53: 334–44.

Walter, A.I., Helgenberger, S., Wiek, A. and Scholz, R.W. (2007) 'Measuring societal effects of transdisciplinary research projects: Design and application of an evaluation method', *Evaluation and Program Planning*, 30: 325–38.

Weiss, C.H. (1997) 'How can theory-based evaluation make greater headway?', *Evaluation Review*, 21: 501–24.

8 Critical issues and challenges for co-producing knowledge for urban change

Merritt Polk

The specific aim of this book is to help academic researchers, urban planners, and policy-makers work together to integrate different types and sources of knowledge in urban areas, thereby making themselves better equipped to solve urban problems. As outlined in Chapter 1, co-production is a way to combine different types of knowledge about urban planning, living, and research to increase the social relevance of the knowledge produced. Various types of co-production were described, including those initiated by the state, social movements, researchers and business. However the type in focus here is what we call researcher-initiated co-production. A number of reasons were given for this focus. Perhaps most important is the need to break the dominance of the linear model over more participatory and inclusive science–policy relationships in order to better address the complex and contested nature of sustainability, the hegemony of neoliberal governance and the current segmentation of policy-making and planning. Mirroring specific social and historical contexts, the examples outlined in the city chapters include a variety of methods, activities, and projects for capturing the values, preferences, experiences, and know-how that are needed to formulate more relevant and legitimate research approaches to urban problem solving in each context. This chapter will focus on what we learned, the problems we encountered, and what areas within both practice and research we need to develop further. It builds upon both the five city chapters (Chapters 2–6) and Chapter 7, as well as on input from researchers and practitioners, including the co-authors of this anthology and other participants from Mistra Urban Futures.

A further objective of this book has been to open up a wider discussion on the urban problematic by bringing together a variety of voices and approaches on researcher initiated knowledge co-production. This was done to create a cosmopolitan community of learning around ways of co-producing knowledge about sustainable urban futures that is based on experiences from the global South and North, from both practitioner and researcher perspectives. In this sense, co-production takes on a wider meaning that includes learning and knowledge exchange across the diverse local contexts and approaches covered in this book. This final chapter will attempt to capture this "meta" view by comparing and synthesizing the diverse approaches and voices on urban knowledge collected here. It begins by summarizing the added value and successes in the city cases. The second section outlines the many critical issues and problems that arose in relation to the problematic nature of many co-production processes and the limitations of their ability to contribute to societal change. Building upon these issues, the final section outlines suggestions for topics that need further scrutiny and attention in co-producing knowledge for sustainable urban futures.

The added value of co-producing knowledge for urban change

In each city example, the participants have developed methods and approaches for co-production that match their specific historical conditions and urban development challenges, from seminars with practitioners and researchers in the CityLabs in Cape Town, to joint ownership and co-funding of projects in Gothenburg. The added values experienced by the participants and extrapolated from the examples given here therefore vary greatly between these different contexts and approaches. Overall, four main areas capture the key added values from knowledge co-production that can be identified in these cities. These are the development of new relationships between practice and research, a sharing of responsibility for knowledge production in urban areas, increased local capacity for research, and increased understanding and insight regarding urban sustainability challenges.

Developing new relationships between practice and research

The first added value of co-production is the creation of spaces and interfaces that better negotiate some of the boundaries between different societal spheres of activity. In these city examples, such spaces and interfaces have enabled the creation of intrinsically new relationships between diverse urban stakeholders and researchers. This has created what can be referred to as "communities of co-production," or spaces for different types of interactive, participatory, and transdisciplinary research. In Gothenburg, for example, prior to the focus on knowledge co-production, researcher–practitioner relationships were predominantly based on individual projects in more traditional consultancy and applied research cooperation, with the city commissioning researchers to work on solving specific, well-delineated, often technical problems, or researchers applying extractive methods of empirical data collection. Through the focus on co-production, these traditional practitioner–researcher relationships have been complemented by more dialogical, interdisciplinary, long-term, and two-way relationships between the political, administrative, and research organizations in the region. The work with co-production in Cape Town has also created a new way for the city and university to work together. The embedding of researchers in city government through the knowledge-transfer program has led to the forming of in-depth and long-term relationships between individuals as well as between city offices and the university.

In Greater Manchester, prior to the work at the Center, relationships between research and practice were based around funder–client–consultant-type arrangements, similar to those in Gothenburg, or around researchers serving as "critical friends." Through the support of the Center, more in-depth and long-term partnerships and new ways of working together across researcher–practitioner boundaries have been developed to populate the "missing middle" of practice–research relationships, through, for instance, working with practitioner-researchers in the Greater Manchester Low Carbon Hub. In addition to creating new two-way relationships between local authorities and universities, many of the co-production practices (for example in Kisumu and some of the CityLabs in Cape Town) have worked on developing multi-way relationships between a diverse range of stakeholders. As with Gothenburg and Manchester, in Cape Town the university had (and continues to have) consultant–client and "critical friend" relationships with the local authority. In Southeast Asia, RMIT's focus on promoting sustainability as a campus-wide strategy through engaged scholarship has also led to more in-depth partnering between the university, researchers, and surrounding communities.

The importance of the need for new relationships between different urban actors is nowhere more clear than in Kisumu. Here, integrating diverse sources of knowledge, expertise, and experiences and building more in-depth collaborative processes enabled the local authorities to better address the severe governance deficiencies regarding service provision, ethnic conflicts, violence, and extreme inequality and poverty. The different types of dialogue and participatory processes in Kisumu provided a degree of increased accountability, helped improve the acceptance of city plans by civil and informal sector groups, and facilitated the implementation of various municipal reforms and activities. Even so, such processes are still viewed with suspicion and are plagued by continued abuses of power by local authorities. As noted in the Gothenburg chapter, the open and participatory "spirit of cooperation" label, which was an important basis for new relationships between practice and research, was also connected to elitist planning processes that promoted land-developer and business interests and systematically excluded civil society and residents from any significant influence. Thus while state- and researcher-initiated co-production has created new relationships for knowledge production in both Gothenburg and Kisumu, it is not surprising to find that well-established power constellations still exist among the ruling elite and special interest groups, and continue to exacerbate ethnic segregation and promote a pro-growth agenda at the expense of social and environmental priorities.

Sharing responsibility for knowledge production

The second added value of co-production is that the new relationships noted above have both promoted and enabled a greater sharing of responsibility for knowledge production between researchers and different practitioner groups. While the methods and approaches described in this book vary greatly, they all show the need for opening up policy-making and governance for non-linear, more interactive relationships with scientific knowledge production and input from civil society, and opening up scientific institutions for a wider influence from practice-based actors. In many of the examples from Gothenburg and Greater Manchester, stakeholders increasingly share responsibility for problem identification and formulation. In both cities, practitioners and researchers were more committed to the research projects and thereby more jointly accountable for the implementation of the results. Rather than seeing researchers as providing the "answer" to public policy challenges, or seeing practitioners as solely responsible for applying research results in policy-making and planning, participants share the task of identifying target audiences and dissemination channels, and translating and packaging results. In Gothenburg, for example, the joint ownership of research projects promoted wider dissemination of scientifically informed input to policy-making and planning processes, as well as brought theory-building and empirical research closer to real-life practice. As noted in Chapter 2, while communicating and implementing project results is still one of the most time-consuming and unmanageable tasks in Gothenburg, the Consortium partners are increasingly sharing this burden.

The move away from the linear model of science–policy interactions has also changed how practitioners and researchers see their respective roles in contributing to different types of policy and urban development processes. Consultancy relationships, where knowledge is commissioned and supplied, have been complemented by roles based on joint responsibility for problem solving, in which scientific knowledge is combined with practice-based expertise. In Cape Town, the city has benefited from moving away from its traditional, consultancy relationship with the university, which was typically characterized by a one-directional flow of information and work. In the knowledge transfer program, this has been replaced by reciprocal relationships between policy processes and research. The embedded researchers straddle the divide between research and practice as active members of their respective city offices, with access to high-level discussions and the inner workings of decision-making processes in the city. These embedded relationships enable researchers to gain more knowledge and understanding, and to share the responsibilities and challenges of the city policy processes. In addition, the CityLabs created diverse groupings of stakeholders, which engaged around specific topics of urban sustainability in Cape Town. Similarly, the Kisumu Action Team gained a broader and more legitimate platform for solving acute conflicts by engaging stakeholders to share responsibility for formulating visions, strategies, and solutions. Although serious problems remain, through this focus on co-producing visions and solutions, the stamp of inefficient governance in Kisumu was to a certain extent transformed into a reputation for improved citizen engagement.

Increased local capacity for collaboration and research

The third identified added value is increased local capacity for addressing complex problems. This increased local capacity includes more resources for local research and growing competence of the participants in the various projects and activities. One of the results in Gothenburg is broader and more in-depth organizational and individual networks, where stakeholders and researchers spanning diverse sectors, governing levels, disciplines, and institutional cultures actively work together. Although co-production in Cape Town was designed as an organizational intervention, it also increased the capacity of individuals to perform their professional roles as researchers and practitioners, particularly in the case of the embedded researchers and officials in the knowledge exchange program. Thus, an important part of increasing local capacity takes place on the individual level. In other instances, capacity building was framed in relation to the community level. In the RMIT work in Southeast Asia, for example, increased local capacity was seen in joint community design or "live projects" where students and researchers, in close collaboration with community groups, co-designed solutions for climate change adaptation, youth housing, and disability day care in vulnerable communities in and around Melbourne and in Vietnam.

In Kisumu, the sharing of responsibility for city development processes had additional emancipatory effects and long-term consequences for the involved citizens. Different types of dialogue-based processes have increased the utilization of knowledge deriving from informal sectors and actors. Such processes have had an emancipatory function by providing opportunities for individuals from civil and informal sectors to appreciate their capacity to share knowledge and act as drivers of change. Overall, this increased local capacity in all of the cities includes the building of new spaces for learning and knowledge exchange, where collaborative processes for integrating knowledge and experiences from different sources are more and more seen as a legitimate way to address complex issues and wicked problems.

Understanding and insights regarding urban sustainability challenges

As noted in Chapter 1, the logic behind the focus on researcher-initiated co-production was the need to free knowledge production processes from the sovereignty of science and the linear model of science–policy relations to allow for more inclusive, multi-stakeholder discourses around sustainability. The aim of co-production is to create processes that can combine experiences and expertise from a variety of urban voices to better impact cities, through different types of interventions, visioning, and understandings. The goal was both to increase the quality of knowledge being used for policy and planning, and to increase the inclusiveness and accountability of planning and policy processes in their specific contexts. However, as is widely discussed, as well as being noted in several places in this book, urban knowledge production is inseparable from relationships of power (Flyvbjerg 1998, 2004; Rydin 2007; Watson 2003). While scientific knowledge has always held a superior position in planning and policy-making, business perspectives and interests have grown increasingly influential and legitimate. At the same time experiences and expertise from civil society, the informal sector and, in some instances, even local authorities, continue to be neglected and marginalized (Polk 2011; Watson 2003). The gap between the influence and power given to the "haves" and "have-nots," described so well in the Greater Manchester chapter, are thus exacerbated by the predominance of the status quo or business-as-usual focus on economic growth, financial crises and austerity, and city branding. While co-production in no way can address the magnitude of these issues, the examples given in this book show how broader platforms and processes for knowledge production can unveil some of the finer details of the rationalities that underlie different framings of sustainability, and give deeper insight into how alternative approaches can be used to challenge the status quo of neoliberal governance. The fourth added value from co-production thus includes increased understanding through mutual learning, greater insights regarding the multiple dimensions of urban sustainability challenges, and a better grasp of the concrete barriers to their solutions.

The focus in Gothenburg on having practitioners and researchers share responsibility for initiating and formulating research projects has led to more informed research questions and approaches. Sharing the data collection and analysis of the material in the mixed project teams promoted continuous mutual learning between the involved researchers and practitioners and their respective contexts. This researcher-initiated approach to co-production thus resulted in a deeper reciprocal appreciation for the different types of expertise that exist in the city and at the university, and, like the Cape Town experience, led to a deeper understanding of the involved policy processes and barriers to sustainability. In Cape Town, the embedded researcher program has allowed for a more detailed understanding of policy- and decision-making in the City. This approach to co-production created concrete structures for first understanding, and then working within the multiple rationalities and drivers for knowledge-based urban change that exist in the city. Through their long-term involvement in the work of various city offices, the researchers have seen first-hand how decision-making is performed in the City, and thus can better grasp the rationalities that underpin decision-making and planning. Discussing these findings with City partners through the Project Advisory Committee has corroborated, strengthened, and added depth to the results, and overall the program has led to a better understanding of the inner-workings of local government in Cape Town. The co-production processes in the CityLabs have also expanded the bodies of knowledge (and made them more holistic) on a wide variety of key themes relating to urban sustainability. For example, in Cape Town these themes included: climate change adaptation and mitigation, urban ecology, housing, health, violence, public spaces and densification. In many cases this has had a real impact on policies and practice.

The history of Kisumu provided in Chapter 4 shows the deep roots of the structural poverty, ethnic reward systems, and corruption that have plagued the city throughout its development. As noted above, the different participatory and dialogue-based processes involving the municipal council and the urban actors contributed to increasing the legitimacy of city governance. The introduction of the scientific approach to community interventions, in KAT, coupled with the "street wisdom" of the civil and informal actors, provided opportunities for researchers and practitioners to learn from each other and establish stronger links between theory and practice, and hence allowing for new insights to arise. For example, NGOs have begun to appreciate the role of academia in driving the agenda of civil society.

The description in Chapter 5 of the history of university engagements with public and private agencies in Greater Manchester shows the crucial importance of creating platforms for more reciprocal, open, and committed relationships between research and practice. The interdisciplinary, co-production approach at SURF embodies a commitment to strategies and approaches that can contribute to a shift from a "knowledge economy for the few" to a "knowledge economy for the many." In Greater Manchester, and similar to the experiences of both Cape Town and Gothenburg, more inclusive processes of knowledge production offered concrete mechanisms to first understand, and then work within the multiple rationalities and drivers for knowledge-based urban change that exist in the region. Some of the added values include in-depth understanding of the mechanisms behind how the trend of establishing "science-cities" in collaboration with universities as engines for innovation and economic growth has effectively confined urban sustainability agendas to a pro-growth approach, and marginalized crucial issues connected to social equality and the environment.

Critical issues and problems for knowledge co-production and implementation for urban change

While the previous section highlighted the benefits and achievements of co-production, this section will describe and discuss some of the many critical issues and problems encountered in the different cities, both within the collaborative processes themselves, and regarding the ability of such processes to impact societal change. Despite the breadth of examples given here, a number of common critical issues and problems can be identified. These are related to the temporal, individual, institutional, and epistemological characteristics of co-production processes.

Time demands and limitations

The most important practical problem for all of the examples of co-production presented in this book is lack of time. Creating new ways of interacting, communicating, and learning across diverse groups and organizational levels takes more time than working alone or within more traditionally linear, sector- and consultancy-based relationships. This results in a continuous need to balance efficiency against the depth and degree of participation of different groups. For example, one of the reasons behind the success of the Gothenburg bid was the three years of meetings—prior to the establishment of the Center—with representatives and leaders from the partner organizations. This prep time established trust and reciprocal learning among the representatives, which in turn facilitated commitment of the individual partners to the Center, and greatly contributed to the competitiveness of the bid. Experiences from RMIT show a similar need to spend significant time establishing mutual trust and forming in-depth relationships on which to base partnerships, especially across ethnic and cultural divides. The multicultural contexts of the RMIT work, for example with indigenous cultures, demand adaptation to local ways of conducting discussions, negotiations, and research, all of which require significantly more time than traditional research approaches. In such instances, clear tensions arise between being culturally sensitive and conducting effective and timely research.

The planning of the projects in Gothenburg also greatly underestimated the amount of time it takes to initiate and manage co-production processes. Translating the vision of co-led and mixed project teams into reality has been a continual problem in Gothenburg. While this is due to a number of factors, including organizational mismatches and communication problems, all of the phases of knowledge co-production take more time than traditional types of research and collaboration. This applies to all steps of the process such as creating joint responsibility and commitment to the project, identifying and engaging participants, jointly formulating and designing research topics, jointly collecting and analyzing data, and co-authoring the project results. In Cape Town, given the novelty of the interactions and the build-up of expectations, time management was and is especially crucial for the embedded researchers and officials on exchange with the university. A further insight, from Greater Manchester, concerns the time needed to understand the motivations and assumptions of the participants that are brought to collaborative processes, within both science and policy contexts. SURF managed such issues by creating time and space for explicitly discussing the drivers and motivations of the individual researchers and practitioners, as well as their goals and expectations. An additional and especially critical factor in time management consists of funders' requirements for in-depth collaboration and partnerships between universities and practice-based organizations, as these expectations typically are not accompanied by an understanding of the demands they place on efficiency and the ability to produce results. Overall, time-management issues and problems, both individual and from funders, plagued a majority of the examples of the types of knowledge exchange and co-production described in this book.

Critical issues for individual participants

The second group of critical issues stems from the role of the individual in co-production processes. While the focus of this book is framed around social interactions and relationships between organizations and groups, the concrete examples in the city cases stress the crucial role of individuals in enabling these new relationships, both as facilitators and as agents of change. Given the complexity of the issues and diversity of involved actors, the ability to negotiate co-production processes is highly dependent on the involved individuals' capacity for self-reflection, patience, and flexibility. In Cape Town, for example, the choice of the leaders of the CityLabs and the embedded researchers was critical, as was their "positionality" (their gender, race, class, education, and professional position) and its relationship to the new constellation of individuals. Some of the character traits that proved important included the ability to be intellectually nimble, build bridges, and act as hybrid practitioner-researchers. In Kisumu, while individual initiative was decisive for facilitating and managing various processes of dialogue and participation, choosing the right individual was equally crucial. Here, when individual interests were put first, they compromised the legitimacy and reputation of entire processes of co-production.

These tensions between individual and collective framings were apparent in all of the cities, but were most developed in the work done in Cape Town, in Greater Manchester, and by RMIT in a number of different cities. In Cape Town, this is discussed in terms of questioning or making explicit the normative starting points of the research. The work in South Africa, with its legacy of apartheid, forces issues of normativity to the fore. Considerations of the relationships between power and knowledge, and how these play out in concrete situations, are in focus throughout their chapter. In Greater Manchester, this is expressed in terms of the reflexive researcher, active intermediaries, and the need to manage the dual focus on relevance and excellence. Part of the work described in the examples from Southeast Asia focuses on adapting to and respecting ethnic and cultural differences. As summarized very well in the work at RMIT, engaged scholarship puts additional demands on all of the participants. These include humility, intercultural respect, genuine interest in the knowledge and expertise of others, willingness to engage in active listening to understand issues from other perspectives, and the ability to put aside preconceived solutions. All of these abilities and traits, while perhaps most pronounced where there are large ethnic and cultural differences, apply to the work being carried out in all of the cities.

This focus on normativity and reflexivity also challenges the traditional division of labor between practitioners and researchers and the roles and status positions that participants have held in the different processes. Academic status and scientific knowledge are usually positioned higher than many other professions and forms of knowledge, and are clearly held to be more legitimate than experiences and expertise coming from non-professional groups in civil society and the informal sector. For example, in Kisumu, a number of researchers were not comfortable in a situation where their previously considered "superior" knowledge and status were brought into question. In Gothenburg, a few of the involved researchers had problems acknowledging differences of opinions from practice-based sources and incorporating them into the overall design of the research approach and subsequent analysis. This resulted in project results that were not especially useful for the practitioners participating in the projects. In Greater Manchester, SURF had to discontinue their evaluation of a municipal funding project because the commissioning municipal actors perceived their results as too critical. There are numerous examples from all of the cities where individuals and organizations "talk the talk" but are not able to "walk the walk" of alternative ways of interacting. Thus, while forming new relationships and sharing responsibility for knowledge production were seen as added values of co-production overall, there were also examples where involved individuals and organizations experienced the process as too uncomfortable and avoided it by isolating themselves, stopping their participation, or ending the relationship.

Another crucial barrier on the individual level is a lack of incentives for researchers and practitioners. For researchers, stepping out of traditional means of knowledge production can be detrimental for a career in academia, where reward systems are based on a high output of peer-reviewed articles. The additional time demands inherent in the examples of co-production given here have clearly negative consequences on academic output. Incentives are equally important for various practitioner groups. Lack of capacity in planning offices and overall time pressures and lack of resources within local authorities, public administrations and civil society organizations greatly limit the ability of officials and representatives to participate in projects and activities, as was seen in all of the cities.

Institutional mismatches

The third group of critical issues includes the different conditions under which individuals and organizations can participate in collaborative processes, given their institutional and organizational affiliations. Despite the advantages of new relationships between science and policy spheres, bridging organizational barriers is highly problematic. Differences in the institutional cultures, mandates, and time frames resulted in a number of organizational mismatches and problems. Such institutional differences among the participating organizations proved difficult, and in some cases impossible to reconcile or transform into meaningful conditions for participation. All of the involved cities and organizations experienced some form of institutional barriers and deficiencies in relation to their work. Structuring the Center and Consortium in Gothenburg as a transdisciplinary organization alleviated some of the barriers to co-production, but also made the barriers and difficulties that still remained even more intractable. Many of these barriers were due to the different organizational cultures of the Consortium partners, and to their lack of congruence with the aims and time demands of the projects. Some of the institutional barriers include established norms and rationalities around problem-solving, the segmentation of sector-based organizations, and reward systems that do not promote transdisciplinary work, both within the university and practice-based organizations.

In Cape Town, for example, agreeing upon the specific topics to be the focus of the embedded researcher program required months of intense negotiations. In addition to long negotiations about the topics/selection of embedded researchers, there were even longer and more complex negotiations about the memorandum of understanding between the university and local authority about the contractual relationship. Both the university and the City are large organizations with complicated bureaucratic procedures that require a great deal of time for negotiations. Such time demands, as noted above, can be detrimental when decisions need to be made quickly, and undermine motivation to participate in new types of relationships. In some of the cities, like Greater Manchester, one of the main institutional barriers to different types of collaboration was a lack of incentives for universities to participate in engaged urban scholarship, which resulted in competition between universities to "win" at innovation–knowledge exchange. One of the most pronounced difficulties in Gothenburg is negotiating the size, interests, and organization of the partners when attempting to both initiate projects and reintegrate results and insights back into their respective research and practice contexts.

All of the cities where co-production was based on relationships between local authorities and universities experienced problems with the different "rhythms" of the participating organizations, as was described in the Cape Town chapter. These differences in conditions include the rhythms of work and time pressures, as well as the autonomy of officials and researchers in their respective agencies and organizations. In Gothenburg, the main institutional barriers included the different time frames of decision-making in the city versus the individuality and relative autonomy of researchers. Within universities, the pressures on researchers, while influenced by teaching workloads and the need to publish and win grants, are much more individually managed. In Cape Town and Gothenburg, this manifested itself in sporadic periods of intense work, common in public organizations, which greatly limited the ability of civil servants to participate in joint processes outside of their official responsibilities. Involving different levels of government, as in Cape Town, Gothenburg and Greater Manchester in partnerships, also leads to additional complications. For example in Cape Town, while the ACC works with both the City and the Provincial Government, they usually do so separately. This is due to the fact that these different levels of governance have a complex relationship that makes it difficult for them to collaborate. On a few occasions, the ACC has even been able to facilitate engagement between the two.

Despite the focus on bridging boundaries, identifying commonalities, and establishing consensus in many of the examples of knowledge co-production presented in this book, collaboration is valued because it is based on differences. Co-production is therefore not only about collapsing roles and functions. As the Greater Manchester chapter illustrates, it "calls our attention to who is doing what and why in the knowledge exchange process". The diversity of individuals and organizations, and their institutional and professional integrity—including the distinctiveness of universities and research—must also remain intact. An important lesson learned is the need to balance creating a base of trust on commonalities between diverse groups with their differences. Increased awareness of what individuals and groups can learn from one another is based not on what they have in common, but on their institutional and individual differences. Thus, as groups develop relationships, they must also respect and nurture the differences that make the collaboration worthwhile. This must be balanced with the reality that differences are also a source of conflict and misunderstanding, and can act as insurmountable barriers to collaboration.

Communication barriers

The fourth group of critical issues is related to the difficulties in communicating across the diverse institutions and groups that are involved in knowledge co-production. Civil society, the informal sector, researchers, and civil servants all comprise multiple rationalities, worldviews, and types and sources of knowledge and experiences. As noted in the introductory chapter, knowledge means different things for these different groups, and different things for the individuals who make up these groups. Sharing responsibility for knowledge production thus foregrounds the tensions that exist between the multiplicity of views of what knowledge means and represents for individuals and collectives, and the diverse ways in which knowledge is created and used in practice and research contexts. Some of these issues have already been discussed, such as the need for self-reflexivity and scrutiny of normative positions. We have also noted examples of the inability of different organizations and individuals to incorporate critical voices and understandings into how they understand and rationalize their work. Many such problems can be explained by the inability of individuals to communicate across different institutional cultures and ways of working, but others are based upon deeply rooted assumptions about what knowledge means—assumptions that are associated with the diverse roles of researchers, decision makers, civil servants, and urban residents. Other more practical problems in communication are simply due to the use of different vocabularies and terminology within researcher and practitioner groups. In the CityLabs there were numerous examples where practitioners genuinely did not understand what the researchers were saying or where social scientists did not understand what the health scientists were talking about. It is also important not to underestimate the communication barriers among researchers and among practitioners, as these can be as profound as those between the researcher-practitioner divide.

More specifically, such communication tensions are based on differences in what is perceived as legitimate knowledge, and what are considered legitimate ways of talking about, producing and using knowledge. Such epistemological differences act as significant barriers to communication in co-production processes. They also have profound consequences for the ability of co-production processes to integrate different types and sources of knowledge, and for how this integration should take place. Such differences are perhaps best exemplified by the tensions between scientific excellence and social relevance, as discussed in the Greater Manchester chapter, and in Chapter 7. One source of this tension is the need to manage the role of non-scientific expertise and knowledge in co-production processes. Within transdisciplinary approaches, the participation and collaboration of practice-based stakeholders is seen as essential to producing usable knowledge. Integrating diverse sources of knowledge is thus considered necessary for adequately grasping the complexity of the urban problematic. However, within other research approaches it is problematic to assume that scientific knowledge can be socially relevant without seriously compromising its reliability and validity (Weingart 2008). The production of scientifically reliable knowledge thus has to be continually balanced against and reconciled with co-creating socially robust and usable knowledge.

A number of risks were also identified regarding socially robust and usable knowledge, such as the risk of being co-opted to produce "policy-based evidence," as in the case of SURF's evaluation of a large city–regional project, or of academic knowledge being used to legitimize the status quo, rather than challenging it or suggesting alternatives. Blurring the boundaries between the science and policy spheres to promote collaboration and knowledge exchange can lead to a loss of critical distance and scientific depth. In some cases, this resulted in the inappropriateness of criticizing close partners, such as the operation of the researchers as "critical friend," in Manchester, and problems with questioning the objectivity of the embedded researchers in their in-between, insider–outsider roles, in Cape Town. Thus, there is a continual need for a reflexive orientation to guard against bias and ensure the quality and validity of scientific knowledge claims. Even if the urban co-production agenda promoted in this book is based upon incorporating intrinsically different drivers and motivations for urban development, it still risks co-option and replication of the narrow economic instrumentalism of the past. Thus knowledge co-production processes must balance on the thin line between consensus and antagonism, focusing, as Chantelle Mouffe would say, on agonism, the ability to work within and recognize the legitimacy of conflicting experiences and views (Mouffe 2005). Applied to the topic here, this means recognizing the legitimacy of both different sources of expertise and ways of producing reliable and usable knowledge.

Topics for future research

The breadth of the substantive issues, urban contexts, and approaches to research-initiated co-production covered in this book makes for an almost unlimited supply of topics deserving of future research. Here we will conclude with two areas that build upon both the overall barriers to sustainability outlined in the introductory chapter and the specific critical issues and problems related to co-production outlined in these conclusions. These will be discussed in terms of the new roles for science and practice that researcher-initiated co-production implies and the need to ground researcher-initiated co-production more strongly within wider approaches to urban planning and societal change in the global South and North.

New roles for science and practice

The shift from a linear model of science–policy relations into one in which science no longer has sole control over the goals and processes of research has profound implications for individual researchers, university and practice-based norms and funding streams. While there have always been different types of collaboration between research and contexts of application, the distinguishing jump with researcher initiated co-production is the quality and depth of the relationships that are now being developed. As noted in Chapter 7, transdisciplinary centers engaged in research that targets societal problem-solving move into new sites and create new types of relationships between universities and different societal actors. This "new production of knowledge," "post-normal science" or "Mode 2," is nothing new in the sense of the longstanding discussions around the third mission, engaged scholarship or action research (Frodeman *et al.* 2010; Gibbons *et al.* 1994; Nowotny *et al.* 2001). However, what is new is the focus on sharing responsibility for knowledge production that occurs within a research context (Maasen and Lieven 2006). What does this mean in practice? One of the most important conclusions that can be drawn from the work presented here is the need to design educational approaches, institutional support, participatory methods, symmetrical processes and evaluation procedures that can better support this repositioning of science in relationship to a more joint responsibility for research that is directed at societal problem solving. One of the main barriers to knowledge co-production is the continuing influence of attitudes, structures and norms which are based on the linear model of science–policy relations within both researcher- and practice-based perspectives and institutions.

First, all of the concrete examples of researcher initiated co-production described in this book point to the important role of individuals as mediators, facilitators, communicators and negotiators between different spheres of activity, ways of creating knowledge, experiences, and ways of thinking. A number of studies bring up the importance of the role of the individual researcher in comparable co-production processes or transdisciplinary research (Bovaird 2007; Lang *et al.* 2012; Pohl *et al.* 2010; Wiek *et al.* 2012). Christian Pohl and colleagues, for example, identify three main roles for researchers in co-production within natural resource management in Kenya, Switzerland, Bolivia and Nepal (Pohl *et al.* 2010). In the first, the *reflective* scientist provides scientific expertise and validates other types of knowledge.

The second, or *intermediary* negotiates between different "thought styles" (ways in which people understand, communicate and gain knowledge about a certain area of activity) and links them to common themes (Pohl *et al.* 2010). The third, the *facilitator*, enhances communication between different "thought collectives" (group of individuals who belong to a specific thought style) and promotes common understanding through joint reflection (Pohl *et al.* 2010: 277). These roles are linked to the need to develop training within higher education that can support the attainment of such skills, for example in handling a diversity of interpretations and experiences of the same phenomena, in understanding underlying power relations, in integrating diverse interests, and in facilitating collective learning processes (Pohl *et al.* 2010: 279).

The examples of co-production presented in this book confirm the need for such roles and training, but also point to the importance of scrutinizing the normative assumptions that underlie all approaches to societal problem-solving, in both practice and research. We also identify an additional need to develop training programs that can increase the skills of not only researchers, but also of civil servants as well as representatives from civil society, to be able to better position their assumptions, expertise and knowledge within wider research and policy contexts. It is clear from the examples given here, as well as from other research, that there is a need to (re)educate individual researchers and practitioners in recognizing how what they bring to the table effects their openness to alternative perspectives and understandings and their ability to be able to co-produce knowledge (May with Perry 2011; Wiek *et al.* 2012). We also need a more in-depth analysis of what implications the focus on normativity and reflexivity has for scientific methodology.

Second, one of the most important issues that is continually noted in empirical studies of different types of transdisciplinary research is the lack of institutional support (Blackstock and Carter 2007; Robinson 2008; Russell *et al.* 2008; Talwar *et al.* 2011). All of the examples of co-production given in this book originated from institutional and practice-based settings that had to position themselves in relation to one-way, linear relationships based upon traditional approaches to policy analysis, extractive research and consultation. The most profound barriers were institutional, both structural and norm based. What is needed is a better understanding of what co-production or transdisciplinary research might entail regarding specific institutional support and adaptation. Mistra Urban Futures itself is an example of such an attempt. However, even given the type of innovative ownership and co-funding that the Gothenburg Consortium represents, the long-term investment, and the research–practice partnerships in the different cities, all of the activities faced huge challenges in fitting into the institutional structures that prevail within their respective universities and within urban policy and planning. One lesson learned is how a smaller scale of more focused and limited forms of in-depth collaboration, as exemplified in the embedded researcher program in Cape Town, was equally or in some ways more manageable and influential than the more ambitious larger scale projects in Gothenburg. We therefore need to rethink issues of scale and what types of institutional arrangements might be the most well suited to manage and assimilate broader, more inclusive learning processes. Finally, and perhaps most importantly, institutional change cannot occur without clear incentives. New forms of funding streams are mandatory as incentives for undertaking this type of highly time-consuming, often intangible work. As noted by RMIT, there is a need to find ways of creating and supporting more dialogical, mutually supportive relationships between universities and their urban settings, in ways that allow the boundaries to be blurred between knowledge creators (the university) and knowledge users (the city) in knowledge co-production.

Third, and closely connected to the last, sharing responsibility for knowledge production entails new ways of working together. One of the reasons that co-production has no clear methodological framework is that it, like all other examples of research attempting to impact societal change, is problem driven. As John Robinson outlines in an article on issue-driven interdisciplinarity, and similar to Flyvbjerg's phronetic planning research, problem-driven research can have no predefined methods or approaches, but must grow out of the methodological considerations that a specific constellation of actors, challenges and historical and social contexts create (Flyvbjerg 2004; Robinson 2008). While the urban development problems met with in these diverse contexts do have many things in common regarding their complexity and wickedness, when it comes to their specific methods or approaches, there are no generic suggestions that can be made. One size does not fit anyone. However, one thing these examples do have in common is that they all represent different types of *experimental* approaches to integrating different types of expertise and knowledge that are based on learning by doing and reflecting. The knowledge transfer program, for example in Cape Town, created an experimental space that was able to flourish because it was under the political radar, and developed slowly over a long period of time. The community based "live projects" at RMIT created experimental spaces and processes for students, teachers and the involved communities to engage in local problem solving together.

All of the methodological approaches presented here also share that they are guided in different ways by ethical issues related both to the distribution of power in urban areas and the moral imperatives of sustainability such as equity, conservation, democracy, and appropriate development. Many of the chapters have discussed the importance of normativity and reflexivity as crucial components of co-production processes. Normativity can be problematic in research because science bases its credibility on being unbiased. The focus on normativity that is raised in this book instead promotes a view that posits that all science is normative to some extent, that all of our positions as officials, civil activists or researchers are based on some form of bias from our experiences, disciplines and worldviews. Researcher-initiated co-production, as well as different forms of transdisciplinary research, sees such "bias" as a strength, as something which needs to be articulated and recognized as important for our ability to learn and communicate with one another. Given this focus on normativity, an important topic for future research is the implications of the tensions between scientific excellence and social relevance for co-production processes and outcomes.

Fourth, and as discussed in Chapter 7, transdisciplinary centers engaged in research that focuses on societal problem-solving do not fit into traditional evaluation approaches that are based on gauging disciplinary or interdisciplinary scientific excellence and impact. They face a number of organizational and management challenges on top of their goals of creating both scientifically excellent, and usable and relevant knowledge. Some of these include addressing and reconciling the needs for communication, performance monitoring, leadership, learning and legitimacy that exist in both practice-based and research spheres. There is thus a need for developing evaluation frameworks that can grasp such management challenges and suitably recognize and assess academic and practice-based impacts. The example of the action-value framework presented in Chapter 7 is one such example. It shows how such management and impact assessment can build upon the perceptions of participant practitioners and researchers, and thereby be situated in a specific urban development context. What remains to be done is to further develop such formative, as well as summative, quality management and evaluation frameworks which can integrate the complexity of both governing a transdisciplinary center and capturing societal impacts. This includes the development of new assessment processes and criteria for evaluating such process, as well as new mechanisms for identifying impacts and effects that can better match the knowledge needs that cities are currently facing.

The potential contributions of research-initiated co-production

As we can see in the different examples of co-production outlined in this book, while the relationships between research and universities, and their counterparts in practitioners and public organizations vary greatly, the overall driving forces for urban development and rationalities behind political strategies for growth are similar. The "millennium", "knowledge" or "culture" city branding in Kisumu, Cape Town, Greater Manchester, and Gothenburg, are all driven by neoliberal mechanisms, though in different guises. While segregation and social inequalities are prevalent in all of the cities, in none are issues of social justice a driving force for urban change. Perhaps only in Kisumu—where social unrest makes governing the city impossible without involving disenfranchised groups in planning—have issues relating to social justice been forced upon municipal and political actors. Social justice is clearly a goal in all of the examples, but it is not where money and effort are most visible.

Given this backdrop for urban sustainability, the final topic for future research is the need to link the experiences and results from researcher-initiated co-production more tightly to theories of urban planning and societal change. This includes the need for revisiting theories within urban planning, interactive research, organizational learning and institutional capacity building with the new experiences of co-production from the different urban development contexts outlined in this book. As numerous researchers from the South have pointed out, theoretical traditions based on Northern case studies from developed countries are not an adequate base for understanding or theorizing the issues that are faced in a majority of urban contexts around the world (Parnell and Robinson 2012; Robinson 2002). There is a need for new approaches to urban theory that are able to incorporate a diversity of urban research experiences from across the globe (Pieterse 2011; Robinson 2002).

Most recently, Watson (2014) focuses on how the importance of the social movement-initiated co-production literature from the South can contribute greatly to more contextualized theory development in urban planning. The examples in this book also suggest that researcher-initiated co-production from the South can have similar contributions in developing methods and theories within sustainability science, transdisciplinary research, and other examples of participatory and interactive research. They also show the importance of the mutual learning that occurs across diverse urban development contexts, not in exchanges of best practice or generic results, but instead in how challenges and insights from one context can inspire new ideas and shift the depth of understanding in another. This is especially clear in the need to develop the methodological issues that have been raised around the roles of reflexivity and normativity in research and practice. Closely related to this, and as noted in the Melbourne chapter, there is both a need to reposition and consolidate the field of urban studies as a transdisciplinary endeavor, and to consider the methodological and theoretical implications of this for the wider field of urban research. Other areas where researcher-initiated co-production could be developed are in relation to social and cultural theories on societal change, such as institutional theory, transition theory, institutional capacity building and planning research (Grin *et al.* 2010; Flyvbjerg 1998, 2004; Healey 2014; Innes and Booher 1999; König 2013; Miller *et al.* 2014; Pierre 2011; Rydin 2007; Wiek *et al.* 2014). Questions deserving of future exploration include:

- What is the potential for different types of co-production to contribute to changing the institutional norms, assumptions, and rules of play that guide policy-making and planning?
- What types of reflexivity are specifically needed for understanding the particular processes and outcomes of researcher-initiated co-production, which include considerations of the normativity of knowledge production and societal change?
- What skills do practitioners and researchers need to populate the "missing middle" and to engage with new forms of university–practice collaboration?

- How can we facilitate learning about the processes or the "how" of knowledge co-production?
- What institutional conditions need to be fulfilled for successful researcher-initiated co-production?

This book has focused on how researchers and practitioners can work together to co-produce knowledge to contribute to solving the complex challenges of attaining sustainable cities. We have given a number of examples of how this was achieved. We will end with a question: What are the potential contributions of researcher-initiated co-production to societal change? A framework for researcher-initiated co-production could provide a wider opportunity for greater South–North exchanges, both empirical and theoretical. For students, researchers and practitioners, co-production could provide an avenue for exploring ideas in more experimental and grounded settings outside of the strict confines of traditional approaches to research and practice. For civil servants and activists, co-production could create more accountability and legitimate platforms for exchanging expertise and experiences. An increasingly educated, empowered and connected populace, also points to the potential that researcher-initiated co-production, as well as the other forms of co-production discussed and mentioned in this book, can have for our ability to create a more grounded basis for sustainable cities in the future. At its best, co-production "walks the walk" of sustainable development by embodying a more inclusive, participatory, and accountable approach to urban research.

References

Blackstock, K.L. and Carter, C.E. (2007) 'Operationalising sustainability science for a sustainability directive? Reflecting on three pilot projects', *Geographical Journal*, 173(4): 343–57.

Bovaird, T. (2007) 'Beyond engagement and participation: User and community co-production of public services', *Public Administration Review*, 67: 846–60.

Flyvbjerg, B. (1998) *Rationality and Power: Democracy in Practice.* Chicago, IL: University of Chicago Press.

(2004) 'Phronetic planning research: Theoretical and methodological reflections', *Planning, Theory & Practice*, 5(3): 283–306.

Foucault, M. (1980) 'Truth and Power', in C. Gordon (ed.), *Power/knowledge: Selected Interviews and Other Writings, 1972–1977*. Brighton: Harvester Press, pp. 109–33.

Frodeman, R. (ed.) (2010) *The Oxford handbook of interdisciplinarity.* Oxford: Oxford University Press.

Gibbons, M., Limoges, C., Nowotny, H., Schwartzman, S., Scott, P. and Trow, M. (1994) *The New Production of Knowledge: The Dynamics of Science and Research in Contemporary Society.* London: Sage Publications.

Grin, J., Rotmans, J. and Schot, J. (2010) *Transitions to Sustainable Development: New Directions in the Study of Long Term Transformative Change.* New York: Routledge

Healey, P. (2014) 'The universal and the contingent: Some reflections on the transnational flow of planning ideas and practices', *Planning Theory*, 11(2): 188–207.

Innes, J. E. and Booher, D. E. (1999) 'Consensus building and complex adaptive systems: a framework for evaluating collaborative planning', *Journal of the American Planning Association*, 65(4): 412–23.
König, A. (ed.) (2013) *Regenerative Sustainable Development of Universities and Cities: The Role of Living Laboratories.* Cheltenham: Edward Elgar.
Lang, D.J., Wiek, A., Bergmann, M., Stauffacher, M., Martens, P., Moll, P., Swilling, M. and Thomas, C.J. (2012) 'Transdisciplinary research in sustainability science: practice, principles and challenges', *Sustainability Science*, 7(Supplement 1): 25–43.
Maasen, S. and Lieven, O. (2006) 'Transdisciplinarity: a new mode of governing science?' *Science and Public Policy*, 33(6): 399–410.
May, T. with Perry, B. (2011) *Social Research and Reflexivity: Context, Consequence and Context*. London: Sage Publications
Miller, T., Wiek, A., Sarewitz, D., Robinson, J., Olsson, L., Kriebel, D. and Loorbach, D. (2014) 'The future of sustainability science: A solutions-oriented research agenda', *Sustainability Science*, 9: 239–46.
Nowotny, H., Scott, P. and Gibbons, M. (2001) *Re-thinking Science: Knowledge and the Public in an Age of Uncertainty.* Cambridge: Polity Press.
Parnell, S. and Robinson, J. (2012) '(Re)theorizing cities from the global south: Looking beyond neoliberalism', *Urban Geography*, 33(4): 539–617.
Pierre, J. (2011) *The Politics of Urban Governance.* London: Palgrave Macmillan.
Pieterse, E. (2011) 'Recasting urban sustainability in the South', *Development*, 54(3): 309–16.
Pohl, C., Rist, S., Zimmermann, A., Fry, P., Gurung, G.S., Schneider, F., Speranza, C.I., Kiteme, B., Boillat, S., Serrano, E., Hadorn, G.H. and Wiesmann, U. (2010) 'Researchers roles in knowledge co-production: experience from sustainability research in Kenya, Switzerland, Bolivia and Nepal', *Science and Public Policy*, 37 (4): 267–81.
Polk, M. (2011) 'Institutional capacity building in urban planning and policy making for sustainable development: success or failure?', *Planning Practice & Research*, 26(2): 185–206.
Robinson, J. (2002) 'Global and world cities: A view from off the map', *International Journal of Urban and Regional Research*, 26(3): 531–54.
——(2008) 'Being undisciplined: Transgressions and intersections in academia and beyond', *Futures*, 40: 70–86.
——(2013) 'The urban now: Theorizing cities beyond the new', *European Journal of Cultural Studies*, 16(6): 659–77.
Russell, A.W., Wickson, F. and Carew, A.L. (2008) 'Transdisciplinarity: Context, contradictions and capacity', *Futures*, 40: 460–72.
Rydin, Y. (2007) 'Re-examining the role of knowledge within planning theory', *Planning Theory*, 6: 52–68.
Talwar, S., Wiek, A. and Robinson, J. (2011) 'User engagement in sustainability research', *Science and Public Policy*, 38(5): 379–90.
Watson, V. (2003) 'Conflicting rationalities: Implications for planning theory and ethics', *Planning Theory and Practice*, 4(4): 395–407.
——(2014) 'Co-production and collaboration in planning—The difference', *Planning, Theory & Practice*, 15(1): 62–76.
Wiek, A., Talwar, S., O'Shea, M. and Robinson, J. (2014) 'Toward a methodological scheme for capturing societal effects of participatory sustainability research', *Research Evaluation*, 23: 1–16.

Wiek, A., Ness, B., Schweizer-Ries, P., Band, F.S. and Farioli, F. (2012) 'From complex systems analysis to transformational change: a comparative appraisal of sustainability science projects', *Sustainability Science*, 7(Supplement 1): 5–24.

Weingart, P. (2008) 'How robust is socially robust knowledge?', in M. Carrier, D. Howard and J. A. Kourany (eds), *The Challenge of the Social and the Pressure of Practice. Science and Values Revisited.* Pittsburgh, PA: University of Pittsburgh Press.

Index

For Product Safety Concerns and Information please contact our EU representative GPSR@taylorandfrancis.com
Taylor & Francis Verlag GmbH, Kaufingerstraße 24, 80331 München, Germany

www.ingramcontent.com/pod-product-compliance
Ingram Content Group UK Ltd.
Pitfield, Milton Keynes, MK11 3LW, UK
UKHW011011180926
475701UK00004B/678
9780367599799